THE
PILTDOWN
MAN HOAX

THE
PILTDOWN
MAN HOAX

CASE CLOSED

MILES RUSSELL

Front cover image: Piltdown Man as reconstructed from bone fragments in 1914. (Sussex Archaeological Society)
Back cover image: Charles Dawson in 1895. (Sussex Archaeological Society)

First published 2012

The History Press
The Mill, Brimscombe Port
Stroud, Gloucestershire, GL5 2QG
www.thehistorypress.co.uk

British Library Cataloguing in Publication Data.
A catalogue record for this book is available from the British Library.

ISBN 978 0 7524 8774 8

Typesetting and origination by The History Press
Printed in Great Britain

CONTENTS

PREFACE

Hastings is a picturesque and historic town in East Sussex, on the south-eastern coast of England. Though it has a dramatically situated medieval castle, a pier, fishing boats and novelty Victorian seaside attractions, it is probably most famous today for a battle that took place on 14 October 1066, some 10km to the north-west of the modern town, between the armies of Harold II, Saxon King of England and William the Bastard, Duke of Normandy.

Today, visitors to Hastings are well served by four excellent museums: the Fisherman's Museum, Shipwreck Heritage Centre, the Old Town Museum and, perched high above the modern town, Hastings Museum and Art Gallery. The latter provides a good introduction to some of the more intriguing characters and former residents of the Sussex town, such as Titus Oates (1649–1705), rector of All Saints' Church, Hastings and religious conspiracy theorist; George Bristow (1863–1947) ornithologist, taxidermist and serial fraudster; Robert Noonan (1870–1911), socialist author who, under the pen name Robert Tressell, wrote the politically inspired novel *The Ragged Trousered Philanthropists*; Aleister Crowley (1875–1949), occultist and celebrity mystic, dubbed 'the wickedest man who ever lived'; Archibald Belaney (1888–1938) local-born prankster who reinvented himself as 'Grey Owl'; a pioneering Native American conservationist and John Logie Baird (1888–1946), who, whilst resident in 1923, created the first working television set. Curiously, there is one notable omission from this list of dream-weavers, visionaries, imaginers, charlatans and manipulators of the truth: Charles Dawson (1864–1916).

Dawson, a local solicitor, is arguably the most famous of all former Hastings residents and characters. Discoverer of *Eoanthropus dawsoni* or 'Piltdown Man', the so-called evolutionary 'missing link' between apes and humans, Dawson was an early twentieth-century media celebrity; the most famous amateur scientist of his generation. From a more local perspective, he was also, in 1889, one of the founding fathers of the Hastings and St Leonards Museum Association, the present-day museum containing much of his extensive antiquarian collection amassed over a prolific forty year period. All of which begs the question: 'why is he not represented in the current story of the town?'

The reason for Dawson's omission from the history of Hastings is simple: the town was, and remains, deeply embarrassed by its premier antiquarian. Former residents Archibald Belaney and George Bristow may both have been fraudsters, but, it is often reasoned, Belaney did good work in the field of conservation (and has, after all, had a Hollywood film made of his life) whilst Bristow is often portrayed as nothing more than a 'cheeky trickster'. Dawson, however, is seen as a charlatan, a serial liar and a deceiver who perpetrated one of the most infamous of all frauds upon both the scientific community and, perhaps more damningly, upon the Great British public. For that, even in death, Hastings seems unwilling to forgive him.

Dawson's name has always featured heavily in the story of the internationally famous Piltdown hoax: a series of archaeological frauds which together appeared to comprise evidence of the earliest human. He was, after all, the finder of the first set of bones and was there at all times during the project to recover the evidence. Since the 1950s, when the fraud was first revealed, it has become apparent, through a careful examination of Charles Dawson's finds, collections and publications, that the hoax perpetrated at Piltdown between 1912 and 1916 was, in reality, the pinnacle of Dawson's alternative career in archaeological, historic, anthropological, literary and biological forgery.

In 2003, the author set out the case against Dawson, using evidence amassed from his forty years' worth of truth manipulation, in the book *Piltdown Man: the Secret life of Charles Dawson*. The question of whether Dawson was in anyway involved in the Piltdown controversy can no longer in doubt; he is the only person implicated at every stage of the fraud and the only one with the means, opportunity and, as far as we can ascertain in the absence of surviving personal correspondence (or indeed of a signed confession), motive.

The book before you now, therefore, is less of a 'Whodunit', as that par-ticular question has now been convincingly resolved, but more of a 'Why' or 'Howdunit'. In this, Charles Dawson, the most successful archaeological and antiquarian forger that the world has ever known, takes centre stage, as is his right. We cannot ever truly see into his mind, but we can trace the history of fraudulent behaviour and see, at every stage of his many attested deceptions, why each and every 'discovery' was considered to be neces-sary. What was Dawson's motivation in all this and what, ultimately, was he trying to achieve?

INTRODUCTION

Even today the name 'Piltdown' sends a shiver through the scientific community, for this quiet Sussex village was the site of a dramatic and daring fraud, the fall out from which continues to affect us.

Between 1908 and 1912, the discovery of human skull fragments, an ape-like jaw and crudely worked flints close to Piltdown was hailed by the world's press as the most sensational archaeological find ever: the 'missing link' that conclusively proved Charles Darwin's theory of evolution. Few archaeological discoveries have the capacity to be front-page news twice over; but 'Piltdown Man' is a rare exception. Forty-one years after he first became famous, the 'Earliest Englishman' was once again a major celebrity, for in November 1953 the world discovered that Piltdown Man had never actually existed, the *London Star* declaring him to be part of 'the biggest scientific hoax of the century'. There never had been a 'missing link' preserved in the gravels of Piltdown; the whole discovery had been part of an elaborate and complex archaeological forgery. The 'Earliest Englishman' was nothing more than a cheap fraud.

In 1912, the year of the Piltdown 'discovery' and only two short years before the mechanised hatred of the Great War spewed its filth across Europe, Britain was an island where class and gender boundaries remained apparently secure with a controlling aristocratic elite and a downtrodden, subservient working class. Women did not posses the vote nor any sense of equality. There were few automatic workers' rights and no real sense of natural justice. Health care depended upon wealth and status; social standing being rigidly enforced from birth. Religious and ethnic minorities were, at best, treated with suspicion, at worst they were targeted for persecution. Abroad, Britain had established a financially successful empire suppressing millions across Asia, Africa, America and Australia.

Yet despite all this, Edwardian Britons, especially those in the more educated middle and upper classes, possessed an immense sense of optimism. The provision of electricity, air flight, transatlantic shipping, rail, telecommunication and the motor car all hinted at greater things to come. There was, thanks in part to popular 'science-fiction' authors such as Jules Verne and H.G. Wells, a tangible sense that technological improvement would one day raise the quality of living for all.

The year 1912 also marked a very real turning point in British history; a time which, in retrospect, we can see the brave new world of technological advancement perceptively beginning to unravel. It was the year in which celebrity adventurer Captain Robert Scott lead the ill-fated expedition to the South Pole; the year in which the 'unsinkable' RMS *Titanic* was lost to the North Atlantic with over 1,500 lives. It is also a point at which the major powers of Europe began their inexorable slide towards global conflict: the London Peace Conference between Britain, Germany, Austro-Hungary, Italy, France and Russia failing to resolve the then escalating violence in the Balkans.

Soon, millions would be dead in both the white heat of battle and the flu pandemic that followed in its wake. The combination of war and plague that ravaged the world between 1914 and 1920 would ultimately prove the catalyst for immense social change, but in 1912 there was no real inkling that such a catastrophic alteration to the basic fabric of life was close at hand. To the bulk of the population, things were pretty much how they'd always been and how they always would be. The discovery of Piltdown Man, the much sought-after 'missing-link' between ape and man, confirmed the belief that British science was the envy of the world. It was yet another triumph at the very heart of the empire.

The Man

No one really *knew* Charles Dawson. Those who thought they did, described him in favourable terms: his colleague Sir Arthur Smith Woodward, calling him 'a delightful colleague … always cheerful, hopeful and overflowing with enthusiasm', a man with 'a restless mind … and never satisfied until he had exhausted all means to solve and understand any problem.' Sir Arthur Keith observed that he had been 'a splendid type' whilst Judges in Uckfield, where Dawson had been chief clerk, noted he had been 'quiet and unassuming with pleasantly smiling eyes … a man of romance.' His wife, Helene, simply remembered him as 'the best and kindest man who ever lived'. In life he had many friends and operated within a wide and diverse social circle. The few photographs of him that exist show an amiable fellow with a jaunty moustache and devilishly twinkling eyes. He appears kindly, jovial, and mischievously intelligent.

Charles was born on 11 July 1864 in Preston, Lancashire to Hugh and Marianne (Mary) Dawson, the fourth of five children, the others being Hugh Leyland (b. 1860), Thomas (b. 1861), Constance (b. 1862) and Arthur Trevor (b. 1866). Hugh, by trade a successful cotton-spinner, used his inheritance to finance his dream of becoming a barrister, but he seems to have been dogged by ill health. Eventually, the family moved to St Leonards-on-Sea, then a new suburb of Hastings in Sussex, where the sea air was considered a significant improvement to the industrial atmosphere of Preston. Whether the change in environment improved Hugh's health, we do not know for, in the national census for 1881, aged 44, he was still being described as a 'Barrister (Not in Practice)'. By then the family, together with a professional nurse and four other domestic staff, were occupying a five-storey town house with a view towards the English Channel.

Of Thomas and Constance Dawson we know almost nothing, but the remaining three siblings, Hugh, Arthur Trevor and Charles, conformed to the Victorian family stereotype: one enrolling in the armed services, one joining the Church and the other following his father's career path (in this case the legal profession). Hugh Leyland Dawson was the only one of the three remaining brothers to attend university, graduating from St John's College Cambridge in 1881. Thereafter he was ordained, finally becoming Vicar of Clandown (near Bath) in 1895, where he remained until his death in 1931. Arthur Trevor Dawson enrolled in the Royal Naval College, Dartmouth, later serving as a lieutenant in the navy. Later he became Managing Director and Superintendent of the Ordnance at Metro-Vickers and Maxims Ltd (one of the premier shipbuilding engineering and arms manufacturers of the early twentieth century) and, for his services to the State, was knighted in 1909, aged only 42. During the Great War, he rose to the rank of naval commander and was made a baronet in 1920. He married Louise Grant, sister of Rear Admiral Noel Grant, a hero of the

First World War, and they purchased a second home, Edgwarebury House in Elstree, London. Successful by any standards, the couple moved within an elite social circle, Arthur Trevor being in a position, in 1906, to introduce his elder brother Charles to King Edward VII.

It is tempting to view Arthur Trevor's success as a spur to his elder brother, a mixture of sibling rivalry and jealousy increasingly driving Charles to enhance his own status and social standing. In truth, however, there is no evidence to suggest that Charles and Arthur Trevor's relationship was ever on anything but friendly and mutually supportive terms. Charles may have envied his brother's success, but perhaps, in reality, both sons had simply inherited their father's own apparent desire for self-improvement; betterment, esteem and status position being hard-wired into the Dawson mindset.

In 1881, Charles, then aged 16, was an 'Articled clerk of Law' working for F.A. Langhams, a firm of solicitors based in London with a branch office close to the family home in Hastings. By 1890 he was working with James Langham, in Uckfield, a modest but steadily expanding East Sussex town, becoming a full partner and taking over the practice in 1900. By 1901, Charles had moved to rented accommodation in Uckfield and in 1906, with George Hart as fellow partner, the branch office became 'Dawson Hart and Co. Solicitors', a name that it retains to this day. In Uckfield, Dawson played a not inconsiderable part within local affairs, being clerk to the Urban District Council and clerk to the Uckfield Justices. He was also appointed secretary of the Uckfield Gas Company, solicitor to the Uckfield Building Society and a trustee of the Eastbourne Building Society. In his professional capacity, furthermore, he acted as steward to a number of large and prosperous estates including the Manors of Barkham, Netherall and Camois.

On 21 January 1905, aged 40, the bachelor Charles married Helene L.E. Postlethwaite, a widow with two grown children, Gladys and Francis Joseph (later Captain) Postlethwaite. Helene, of Franco-Irish descent, may have been introduced to Charles through his younger brother Arthur Trevor for, at the time of their engagement, she was resident in Park Lane and a prominent member of Mayfair society. The wedding took place at Christ Church, Mayfair and was by all accounts a rather grand event, the bride being given away by Sir James Joicey MP; Dawson's best man being the wonderfully named Basil Bagshot de la Bere of Buxted. Following the wedding reception, held at Trevor and Louise Dawson's London residence, Charles and his new wife honeymooned in Rome. Afterwards, Dawson made plans to return to Sussex and in 1907 the family moved into Castle Lodge, a spacious town house set in the grounds of a Medieval castle at the centre of the market town of Lewes.

From his earliest days Charles was a fossil collector, gathering specimens from the coast, cliffs and quarries around Hastings, much of which he donated to the British Museum (Natural History Museum). In gratitude, the museum conferred upon Dawson the title 'honorary collector' and in 1885, in recognition of his many varied discoveries, he was elected a Fellow of the Geological Society, quite an achievement for a man who was only then aged 21. In 1889, Dawson co-founded the Hastings and St Leonards Museum Association, one of the first voluntary museum friends' groups established in Britain. At this time Dawson also found himself acting as solicitor to a number of prominent antiquarian collectors in the region and was, therefore, in a good position to catalogue artefacts bequeathed or otherwise donated to Hastings Museum. By 1890, he was even conducting his own excavations in the town, most notably on and around Castle Hill. Unsurprisingly, perhaps, Dawson found himself on the Hastings Museum Committee, in charge of the acquisition of artefacts and historical documents whilst in return the museum proved the ideal place for Dawson to display his ever-expanding collection.

By 1892, Dawson was turning his attention to matters archaeological, joining the Sussex Archaeological Society as honorary local secretary for Uckfield. The following year he was chosen, together with the society's librarian and clerk, to represent the organisation at the fifth Congress of Archaeological Societies in association with the prestigious London based Society of Antiquaries. By 1893 he found the time to co-direct excavations beneath Hastings Castle and in the Lavant Caves near Chichester, with John Lewis. In recognition of his many discoveries, theories and constant hard work in the field of antiquarian research, Dawson was elected a fellow of the Society of Antiquaries London in 1895. At the age of 31, and without a university degree to his name, he was now Charles Dawson FGS, FSA.

In 1901 Dawson was present at the founding of the Sussex Record Society, a sister group to the Archaeological Society intended to research, curate and publish historical documents relating to the county of Sussex. In April of 1902, by the time of the group's first annual general meeting, Dawson had become a full member of the council. At this point he had begun to write extensively (and tirelessly) on all aspects of Sussex history and archaeology. He studied ancient quarries and mines, re-analysed the Bayeux Tapestry and produced the first definite study of Hastings Castle.

He took a great interest in the fledgling art of photography and was invited on a number of archaeological expeditions, most notably in 1907 accompanying John Ray in the exhumation of prehistoric burials in Eastbourne.

By now, Charles was investigating more unorthodox aspects of the natural world, including toads petrified in flint, sea serpents, horned cart horses and a new species of human. It was even reported that in the latter years of his life, he was experimenting with phosphorescent bullets as a deterrent to the Zeppelin attacks on London. As his scientific interests diversified, he began to press his candidature for Fellowship of the Royal Society, something that, as an amateur enthusiast, would have marked the pinnacle of academic recognition. His first candidacy certificate was filed in December 1913, and was renewed every year, without success, until his death in 1916.

His greatest claim to fame, however, was the discovery of the so-called missing link between ape and man, derived from the Sussex village of Piltdown. At the time of its announcement in December 1912, the solicitor from Sussex found himself, together with his colleague Arthur Smith Woodward of the then Natural History Museum at the centre of a worldwide media storm. In honour of its discoverer, the new species of human was given the name *Eoanthropus dawsoni*, literally 'Dawson's man of the dawn'. He was now the most successful antiquarian in Europe. A painting unveiled at the Royal Academy in 1915 shows him at the peak of his academic achievements, standing with the greatest scientific minds of Edwardian England with an image of Charles Darwin, father of evolutionary science, sitting contentedly in the background.

Unfortunately, at the point of this, his greatest achievement, Dawson was taken gravely ill. The exact nature of his sickness remains unknown, though it has been suggested he was suffering from pernicious anaemia. Despite resting for a brief period at home during the spring of 1916, by June his condition had worsened, and he was confined to bed. Early on 10 August 1916, aged only 52, he died. Death robbed him of the chance of knighthood, an honour that many others associated with the Piltdown find were to receive. The funeral, held on 12 August at the Church of St John Sub-Castro in Lewes, was attended by over 100 mourners, the service being conducted in part by his brother, the Reverend Hugh Leyland Dawson.

Twenty two years later, on 22 July 1938, as a lasting tribute to both Dawson and his *Eoanthropus dawsoni*, a monolith was erected at Piltdown, at the spot where the skull had been first been discovered. In September 1950, the Natural History Museum supervised the opening of a section through the adjacent gravel terrace, the edges of which were bricked up with two glass windows set along the western edge to act as a permanent witness section. In 1952 the area was designated as a Geological Reserve and National Monument by the Nature Conservancy. Dawson's legacy was complete.

The Hoax

There the story might have ended, were it not for some persistent and nagging doubts that there was 'something not quite right' about the Piltdown discovery. There never had been full agreement over what the finds actually represented, nor how they could be reconstructed. Worse, as the 1920s and '30s played out, the question of where exactly to place *Eoanthropus dawsoni* within the tree of human evolution could not satisfactorily be resolved. When Dawson had first produced the bones from Sussex, there was little or no comparative material with which to work. Few fossil remains of early humans had been found and there was a great deal of argument surrounding the way in which to interpret lines of descent.

As more fossil discoveries were made, especially during fieldwork conducted in China and Africa, it appeared that the aspects that best defined *Eoanthropus*, a human forehead and an ape-like jaw, were not present elsewhere. In fact, new fossil remains demonstrated that human-like teeth and jaw were a very early feature in human development, whereas the brain and forehead changed more gradually. Piltdown had these key features in reverse; it was an anomaly that scientists were beginning to find incredibly embarrassing.

Analysis of the fluorine content of *Eoanthropus* undertaken in the late 1940s created more problems, suggesting that the bones of 'Dawn man' could not be any more than 50,000 years old. Such a date meant that Piltdown Man could not have been an ancestor of the modern human, merely an archaic form or genetic 'throwback'. Most scientists now felt it easier to simply ignore the old man from Sussex altogether. Thankfully, resolution to the problem came in December 1953 when the myth that *Eoanthropus dawsoni* had been a genuine living being was exploded forever. Joseph Weiner, Kenneth Oakley and Wilfred Le Gros Clark delivered their fatal blow in the pages of the *Bulletin of the British* [Natural History] *Museum Geology*. Entitled simply 'The solution of the Piltdown problem', the article revealed that the jaw and teeth of Piltdown Man had all been forged. Piltdown Man had never lived: he was nothing more than a cheap hoax.

The main question since the exposure has been with regard to the identity of the perpetrator. Weiner, in his 1955 book *The Piltdown Forgery*, was sure that it was 'not possible to maintain that Dawson could not have been the actual perpetrator; he had the ability, the experience, and, whatever we surmise may have been the motive, he was at all material times in a position to pursue the deception throughout its various phases'. The backlash that followed seems, in retrospect, hardly surprising for, nearly forty years after his death, Dawson remained a significant and well-liked figure within the worlds of palaeontology, anthropology and antiquarian archaeology. To attack him meant to assault every single one of the scientific experts that had lined up to support him and bolster his research. Surviving members of the family were livid: his stepson, Captain Postlethwaite writing to the London *Times* in outraged terms within days of the first press release.

Since 1955, the argument has swung both for and against the possibility that Dawson, the discoverer of Piltdown Man, could also have been its creator. Most of the discussions rely on the premise that Piltdown was a 'one off', a single, if rather elaborate hoax, designed to fool the scientific community, embarrass key figures of the Establishment or to verify (or perhaps even to discredit) fledgling models of human evolution. Under such circumstances, one of any number of people may plausibly be held responsible, though the most 'usual' of cited suspects are Charles Dawson, Arthur Smith Woodward, Pierre Teilhard de Chardin and Venus Hargreaves (all members of the original excavation team), as well as the writer Arthur Conan Doyle, the anatomist Arthur Keith, museum curator William Butterfield, zoologist Martin Hinton, palaeontologist William Solas, neurologist Grafton Elliot Smith, jeweller Lewis Abbott and chemists John Hewitt and Samuel Woodhead.

In 2003, the author published the results of research into the career and discoveries of Charles Dawson FGS, FSA. The book, entitled *Piltdown Man: the secret life of Charles Dawson and the world's greatest archaeological hoax* examined all of Charles Dawson's major antiquarian finds, looking at both the composition and circumstances of each, and his many varied publications. Of the discoveries listed, it became apparent that at least thirty-three, including the Piltdown skull, mandible, teeth, animal bone assemblage and flint artefacts, the 'shadow' figures of Hastings Castle, a hafted stone axe from near Eastbourne, an ancient boat from Bexhill, Roman bricks from Pevensey, artefacts from the Lavant Caves near Chichester, a cast-iron Roman statuette from Beauport Park, a Bronze Age antler hammer from Bulverhythe, a Chinese vase and a toad preserved in stone, were all clear fakes. The chief (and in some cases *only*) suspect was none other than Charles Dawson himself.

In addition to these forgeries, it was possible to show that other dubious discoveries, such as a horseshoe from Uckfield, a spur from Lewes and a prehistoric standard mount from St Leonards-on-Sea were fakes, probably generated by Dawson whilst an 'Arabic' anvil and a small axe from Beauport Park were possible fakes associated with the Sussex solicitor. With regard to Dawson's publications, at least one was clearly plagiarised whilst papers on Dene Holes, the Red Hills of Essex and Sussex iron, pottery and glass, together with the two volume study the *History of Hastings Castle* were all compilations of other people's work.

That Charles Dawson was the ultimate designer, creator, instigator and perpetrator of the Piltdown hoax cannot be in doubt. Dawson was the only man with the experience, the knowledge and the 'form' to create the fakes and the various other forgeries that preceded it. He was the only person who had the means, opportunity and motive throughout nearly twenty-five years of archaeological, geological and antiquarian deceit. He was the only one who truly benefitted, not just from Piltdown Man, but from all the many 'finds' that came before it. In short, Piltdown was not a 'one off', more the culmination of a life's work.

It was a dangerous game, but Charles Dawson played it so very well.

PHASE 1

A SPLENDID FELLOW

We do not know exactly at what point Charles Dawson was seduced by the idea of forgery, but it is clear that the first evidence of fraud comes with one of the earliest of his discoveries.

The Fossil

Dawson had been an avid collector of fossils from a very young age, gathering specimens from the cliffs and quarries close to the family home in Hastings. In many of these formative searches, Dawson had been encouraged by Samuel Husbands Beckles, a Fellow of the Royal Society and a distinguished geologist, then in his twilight years. Beckles, who as a lawyer may well have known Dawson's father Hugh, had lived in St Leonards-on-Sea since his retirement in 1845. He spent much of his time exploring the fossiliferous outcrops of Sussex and Hampshire, being credited as the first to recognise dinosaur footprints on the Isle of Wight. He discovered a number of dinosaur species new to science, including a small herbivore, named *Echinodon becklesii* in his honour, and a bipedal carnivore named (much later) as *Becklespinax*. By the time of his death, in 1890, he had amassed a huge collection of fossils, including a significant number retrieved from 'Beckles' Pit', a 600m² excavation into the rocks of Durlston Bay in Dorset that he had overseen throughout 1857.

Together, the two men, Dawson and Beckles, collected an impressive array of fossils, the prize of which, noted as the 'finest extant example' of ganoid (plated) fish *Lepidotus mantelli*, was donated to the Natural History Museum in 1884. Other discoveries followed, including three new species of dinosaur, one of which was named *Iguanodon dawsoni* after the young solicitor by the palaeontologist Richard Lydekker, and a new form of fossil plant, later named *Salaginella dawsoni*. Under Beckles' apprenticeship, the 20-year-old Dawson was getting himself noticed. In gratitude for the burgeoning archive of dinosaur remains, the Natural History Museum conferred upon Dawson the title of 'honorary collector' and in 1885, thanks to his new circle of scientific friends, he was put forward and elected a Fellow of the Geological Society.

There is nothing overtly suspicious about the majority of Dawson's fossil discoveries, in fact specimens such as *Lepidotus*, *Salaginella* and *Iguanodon* are exactly the sort of things that a determined amateur palaeontologist feverishly searching the quarries of East Sussex would, during the latter years of the nineteenth century, be expected to find. Neither is there any doubt concerning the integrity of Samuel Beckles whose collection was donated to the Natural History Museum following his death in 1890. It was in the months following the passing of this great amateur palaeontologist, however, that Dawson's alternative career in deception took its first faltering steps.

As a solicitor with a passion for fossil remains and a close personal friend of the late Samuel Husbands Beckles, Charles Dawson was the perfect choice to sort, catalogue and record the collector's archive prior to its deposition in the Natural History Museum. In fact, Arthur Smith Woodward, then an assistant curator at the museum, observed that the young man 'gave much help to the British [Natural History] Museum in labelling the collection of Wealden fossils which was acquired from that gentleman's executors'. Not all the finds, however, made it to the stores of the London museum, some remaining (largely unlabelled and lacking full documentation) in Dawson's hands, finally ending up in the stores of the Hastings and St Leonards Museum. Dawson, it would seem, was reluctant to hand over all of the Beckles collection. Perhaps he kept some fossils as a keepsake to remind him of his erstwhile friend, or perhaps he felt that, as joint finder, he deserved to retain some of the pieces. Perhaps he believed that the more local Hastings institution deserved some of the collection for the purposes of display and research. Perhaps there was a more sinister motive behind the retention of specific artefacts.

Late in November 1891, Arthur Smith Woodward presented a paper concerning an exciting new fossil discovery before the Zoological Society of London. The find, a single tooth, was potentially earth shattering, as it seemed to provide the first evidence of a 'European Cretaceous Mammal':

an important missing link in the history of life on earth. The tooth had been found, Woodward reported, by 'Mr Charles Dawson of Uckfield, in an irregular mass of communated fish and reptile bones, with scales and teeth' from a quarry near Hastings. Undeterred that the precise location, date and circumstances of the tooth's discovery were vague, Woodward observed that its size, when combined with the shape of the crown, strongly suggested that the fossil had derived from a wholly new species of the mammal order *Multituberculata*. This new species could, Woodward suggested, until the acquisition of further material 'bear the provisional name of *Plagiaulax dawsoni*, in honour of its discoverer'.

The order *Multituberculata* first appeared in the Jurassic period, between 206 and 144 million years ago, being at their most diverse and widespread during the late Cretaceous. Multituberculates do not belong to any of the groups of mammals alive today. They were small and hairy in appearance, their pelvic anatomy suggesting that they gave birth to tiny, marsupial-like young. The final lower premolars of most Multituberculates formed enlarged, serrated blades, such teeth often being described as 'plagiaulacoid' after the Mesozoic Multituberculate genus *Plagiaulax*.

The tooth that Dawson had presented to Woodward was larger than any known example of the genus *Plagiaulax*. It was also far more abraded than was common, Woodward observing the extraordinary amount of wear 'to which the crown has been subjected', having lost nearly all of its enamel. It was the patterning of wear that most seemed to perplex the young geologist, especially as the abrasion had not been produced 'entirely by an upward and downward or antero-posterior motion, of which the jaws of the know Multituberculata seem have been alone capable'. Any doubts concerning the antiquity of the abrasion were dispelled by Woodward, however, who noted that when he had first received it, the fossil had been so firmly embedded within its soil matrix that only long and diligent work by the technicians of the British Museum laboratory could satisfactorily detach it.

We now know, however, that the 'discovery' was in fact nothing of the sort; careful examination of the tooth showing that the side-to-side abrasion sustained, which Woodward noted as being otherwise unknown in the natural wear of this order of ancient mammal, is wholly artificial. Such damage, which had eroded the crown and much of the original enamel, could only have occurred through a programme of extensive and prolonged post-mortem rubbing with an iron file. In short, *Plagiaulax dawsoni* was a fake.

Quite why Dawson moved from 'honorary collector' and supplier of genuine finds to master forger it is impossible to say. Perhaps time spent fossil hunting in the quarries of east Sussex was not producing sufficient rewards; perhaps the process was simply taking too long or was losing its appeal (Dawson's spare time increasingly being taken up with his career and other interests); perhaps he just craved greater academic recognition? For whatever reason, at some point in 1891, Charles Dawson made the decision to gently manipulate existing geological data. He created a fraud.

Doctoring a Plagiaulacoid tooth taken, in all probability, from the extensive collection of his late colleague Samuel Beckles, Dawson filed down the crown, eroding much of the enamel in the process, and, in doing so, manufactured evidence for a wholly new species. Interestingly, although the find was only small, Dawson had hit upon the best way of increasing artefact significance and generating academic interest: he had created a 'missing-link' or transitional form, in this case between 'terrible lizard' and mammal. Transitional forms were later to play a prominent part in his antiquarian career, culminating, of course, with the missing link between ape and man that Piltdown so clearly represented. *Plagiaulax dawsoni*, however, was in retrospect a basic and fairly clumsy fraud, and anyone with access to a microscope and a degree of scepticism could easily have spotted it. At the time of its announcement, though, no one in the scientific community doubted its authenticity, especially as Dawson had gone to extreme lengths to ensure that the tooth was apparently still embedded in soil at the time it was donated to the Natural History Museum.

Dawson had, in the fabrication of wear patterns across the fossil tooth, also established a second vital aspect of establishing credibility, something which would become a key part of his *modus operandi*: the academic dupe. An academic or expert dupe was one who was unaware of a particular fraud, but who was only too happy to witness and verify it (therefore also indirectly verifying the reliabilty of the 'finder'). Establishing an unwitting dupe early on in the 'discovery' process not only helped increase the perceived authenticity of an object, making it more difficult to doubt or discredit, but it also confirmed both identification and attribution in a way that was acceptable to the wider scientific community.

In the case of *Plagiaulax dawsoni*, the solicitor from Sussex had found the first of a number of unwitting (and totally willing) victims: Arthur Smith Woodward. Woodward had no reason to doubt the word of Dawson, but neither did he fully investigate the nature of the 'find' nor interrogate the circumstances of its 'discovery'. His failure to spot the blatantly artificial wear patterns on the surface of the fossil tooth is forgivable, given that Dawson was both an enthusiastic fossil hunter and a man of the law, but his unambiguous acceptance and enthusiastic support of the forgery was something that Dawson would use again and again, to ever-greater success, most notably when it came to the fabrication of Piltdown Man.

Dawson's path to infamy had begun.

The Caves

Plagiaulax dawsoni had created a small amount of interest, further establishing Charles Dawson's credentials as a fossil hunter and amateur scientist of note (and providing another example of a species named after him), but, in the wider academic and antiquarian world, it was still small beer.

In the years that followed 1891, Dawson sporadically continued his fossil hunting expeditions, but prolonged work in the field of palaeontology ultimately left him little scope to improve his academic and scientific standing. He needed to diversify; to set his sights upon new targets; new areas of potential. Shortly after the formal identification of *Plagiaulax dawsoni*, Dawson switched his attention to the antiquities of man.

In 1892 he joined the Sussex Archaeological Society, becoming honorary local secretary for Uckfield, and began his own archaeological examination of Castle Hill in Hastings, the exact details of which remain frustratingly vague. It seems that the results of this phase of work were not quite to Dawson's liking, the finds not up to his high level of expectation. The excavations did, however, provide him with the opportunity to gain valuable fieldwork experience and to hone his craft as an antiquarian. Now he was ready to work upon sites which were more likely to produce results of the sort that he needed to advance his academic standing. In 1893 he commenced the exploration and examination of a series of tunnels, the first of which were the Lavant Caves, near Chichester, followed soon after by the Medieval passageways beneath Hastings Castle.

In 1916, the antiquarian Hadrian Allcroft wrote in angry terms concerning the archaeological investigation of the Lavant Caves, to the north of Chichester in Sussex, singling out Charles Dawson for particular abuse. 'The skill of a north-country miner,' Allcroft raged in the pages of the *Sussex Archaeological Collections*, 'would have dealt easily with the matter at the outset, and enabled the whole area to be cleared, searched and planned. As it is, the Caves, it is to be feared, are now lost for all time, and their secrets with them, while even the few "finds" are difficult of access to the majority'.

The main reason for Allcroft's displeasure was that, two decades after the site had been dug, the results remained unpublished and largely unknown. To make matters worse, the artefacts deriving from the dig had been dispersed whilst the caves themselves had been sealed, due to the threatened collapse of the roof, something that emphatically prevented any further investigation. Attempting to compile what little was known about the site, Allcroft complained bitterly that he had faced 'the greatest of difficulty in ascertaining something of the facts'.

What information we do have today concerning the investigation of the Lavant Caves in 1893 comes from secondary sources, primarily from observations made by visitors to the site. As far as we can tell, Dawson only publically discussed his work at Lavant once, in a paper presented to a meeting of the Sussex Archaeological Society in August 1893. Some details of the lecture were reported shortly after in the *Sussex Daily News* and Dawson himself supplied some detail to George Clinch, who used it in his chapter on 'Early Man' in the first volume of the *Victoria County History of Sussex*.

According to Allcroft, using information culled from Dawson's original lecture, the existence of a network of tunnels at Lavant had for years prior to 1890 been suspected due to the refusal of livestock 'to draw the ploughs over the thin roof of chalk which concealed the Caves'. Confirmation of the presence of underground workings came in around 1890 when an unnamed shepherd lost two hurdles he had been carrying through an opening in the roof of the buried feature. Realising the potential significance of the find, the landowner, the 6th Duke of Richmond, forced a brick and mortar stairway into the caves and commissioned two members of the Sussex Archaeological Society, Charles Dawson and John Lewis, to conduct further investigations.

Quite how and why Dawson got the commission, especially as his work to date had been primarily examining the geology and fossils of the Hastings area, is unclear. By 1893 Dawson was, however the honorary secretary of the Sussex Archaeological Society for Uckfield and he had already represented the organisation at the Congress of Archaeological Societies. Perhaps, as one of the up and coming new members of the society, he was simply the logical (or most available) choice. John Sawyer, commenting on the excavation of the caves in the July 1893 edition of *The Antiquary* notes enigmatically that Dawson had a 'penchant for explorations of this kind', whilst Lewis had apparently 'done some good work in the same line, especially in India'.

Little is really known about Dawson's collaborator in the project, John Lewis. His 1896 certificate of candidature for election to Fellowship of the Society of Antiquaries, following his collaboration with Dawson at Hastings and Lavant, notes his occupation as 'retired CE, formerly in the service of the Indian Government', his qualification being as a result of 'archaeological research in India and England'. The exact nature of this earlier work is unknown though the 'CE' in his application was probably an abbreviation for civil engineer, and it is known that a 'John Lewis' was employed as a permanent way inspector on the civil engineering staff of the North, and East, Bengal State Railways until 1893. It would seem likely, therefore, that the two Lewises were in reality the same, especially as 1893, the year that the Lewis employed by the North, and East, Bengal State Railways retired, was also the same year that the John Lewis 'formerly in the service of the Indian Government' came to work in the Lavant Caves with Charles Dawson. Dawson later thanked Lewis for providing plans of the Uckfield gas borings, so perhaps, in the capacity of 'CE', Lewis had already put his experience on the Indian railway to work in a similar field on his return to England. If he was indeed an experienced civil engineer, then Lewis would certainly have been an invaluable asset to Dawson in his clearance and recording of the Lavant Caves.

The presumption at the start of work was that the Lavant Caves represented a form of prehistoric flint extraction pit. Flint mines were, by the end of the nineteenth century, hot news as early archaeologists struggled to expose the full antiquity of human endeavour. Between 1868 and 1870, Cannon William Greenwell oversaw the first full excavation of a flint mine-shaft in Britain, at Grimes Graves in Norfolk, to a depth of some 12m. The shaft had cut through and exploited three horizontal seams of subterranean flint, classified by Greenwell as the 'topstone', 'wallstone' and 'floorstone'. The 'floorstone' deposits at the base of the shaft represented the good quality flint that was so extensively exploited by Neolithic miners through the cutting of radiating galleries. Greenwell's work in Norfolk kindled huge interest in prehistoric mine sites, especially those detected in Sussex. In 1873 Ernest Willett began work at Cissbury, to the north of Worthing, where he traced a Neolithic shaft to its full depth of 4.2m. After Willett came Plumpton Tindall, Colonel Augustus Lane Fox, J. Park Harrison, Professor

George Rolleston and Sir Alexander Gordon, all of whom opened new areas across the chalk hill of Cissbury between 1873 and 1878.

Grimes Graves and Cissbury had a massive impact upon antiquarian and early archaeological thinking, for they suggested that pretty much any hole in the ground could potentially represent the remains of a substantial prehistoric mine. For a while, shafts of all shape and size, together with a variety of marl pits, chalk quarries, dene holes, wells and natural solution pipes, were grouped together with genuine prehistoric mines, as early archaeologists attempted to define and understand this new class of field monument.

In 1893, having gained access to the caves via the new staircase, Dawson and Lewis set about examining the subterranean workings. Dawson busied himself with the investigation of floor debris, whilst Lewis began recording the nature and extent of the tunnels. It is clear that Dawson and Lewis did not do all the work themselves, and that, as with many large excavations of their day, much of the 'heavy work' (the basic loosening, shifting and extraction of spoil) was conducted by a local labour force, possibly a team drafted in by the Duke of Richmond. Only two of the team are specifically named on the list of artefact labels recorded from Goodwood House, where the finds from the dig were later stored: a Mr Lawrence and 'Workman Hammond', but there would undoubtedly have been more.

The original entrance to the caves could not be found, the Duke of Richmond's staircase, which provided the only point of access for the excavation team, having been inserted through the collapsed roof of the main domed chamber. John Sawyer, clerk of the Sussex Archaeological Society, who visited the caves during the course of 1893, observed that 'the passages are, for the most part, nearly choked with loose chalk, all of which it may be hoped will eventually be cleared away'. A total of five irregular galleries or interconnecting tunnels and three domed chambers were partially cleared during the 1893 season. The maximum height of the galleries were noted as being between 4ft and 5ft (1.2m and 1.5m). The full size of the chambers was unfortunately not recorded, though Mary Wyndham, daughter of the 2nd Lord Leconfield, who visited the site in 1893, noted 'the cave consists of a tunnel 30 yards long, down which you walk doubled up, till you reach a small chamber in which you can stand upright'.

Something akin to 5ft (1.5m) of rubble had built up across the main floor of the cave, Dawson, via Allcroft, noting that this comprised two main layers: 'the lower, upwards of 2ft in thickness, was of finely crumbled and compacted chalk, probably trodden to powder by the feet of those

who used the Caves' the upper levels comprising 'larger and looser fragments of chalk, the accumulation of later falls from the roof at a time when
the Caves were no longer frequented' . This is quite a deposit, and could
possibly indicate the general instability of the cave roof. Alternatively, the
mass of compacted chalk could relate to the deliberate infilling of redun-
dant workspaces by those engaged in the original quarrying. In a number
of Neolithic flint mines, for instance, completed galleries were often filled
with rubble derived from excavations in other areas of the shaft. Such infilling would suggest that the Neolithic miners packed abandoned tunnels
with debris excavated from new areas so as to avoid hauling all material
out to the surface.

As the dig progressed, however, both Dawson and, to some extent, Lewis,
must have realised that the Lavant Caves were not in anyway similar to the
flint mines recently examined at Cissbury a few miles to the east and at
Grimes Graves in Norfolk. It was, quite simply, morphologically distinct,
possessing neither a central shaft nor radiating galleries and, rather damningly, there was no indication of any subterranean flint. Worse, the artefactual
assemblage derived from the investigation was pretty much non-existent;
the upper levels of chalk rubble removed from the cave containing only
a few nondescript flint flakes. Clearly, the cave, although spatially impressive, was not going to set the antiquarian world alight. It was not going
to improve Charles Dawson's standing in the archaeological and academic
community one little bit, a fact which must have frustrated the solicitor
coming so soon after his inconclusive diggings on Castle Hill in Hastings.
Something needed to be done and fast.

Dawson's solution was to introduce a better quality of artefact to the
lower levels of the cave; he was, quite literally, going to 'salt the mine'. As
the excavation team cleared the softer chalk layer lining the bottom of the
subterranean cutting, the quantity of finds began to increase dramatically. To
begin with the assemblage proved confusingly diverse, comprising, amongst
other things, a few pieces of Roman mosaic, Roman bronze ware and pottery, some sixteenth-century metalwork, a Georgian halfpenny and a few
human teeth, together with objects of amber, lead and silver. By the end of
the investigation the flow of these 'discoveries' stabilised with rather more
secure evidence of prehistoric activity, namely more (and better manufactured) flint work, a hollowed-out chalk block (akin to the so-called 'miners'
lamps' from Cissbury) and a tine of red deer antler (antlers being the primary digging tool employed in the Neolithic mines of both Sussex and

Norfolk). All of these artefacts had, given that the Lavant Caves were, in all likelihood, part of a medieval or later chalk quarry, been introduced for the excavation team to find. Crucially, all the finds, which were probably selected from Dawson's own, by now rather extensive, antiquarian collection, were small and could easily be transported into the caves, hidden in coat or trouser pockets, to be securely placed within unexcavated chalk debris that littered the floor. Being the director of operations, Dawson was able to do this without suspicion, for it was his job to oversee all aspects of cave clearance.

Dawson and Lewis could of course have been working in unison, both hoping that by making the Lavant Caves appear more impressive than they really were, they would increase their chances of academic recognition. The relatively small-scale nature of the deception may, however, argue against such a scenario for, if both men were working in concord they would surely have generated something on a far grander scale. All the artefacts were compact enough to be secreted about the person and such illicit transportation of objects would only really be necessary if the person organising the fraud felt that they were under some form of supervision. If both site directors had been working in unison, it would have proved unnecessary to have smuggled anything in, for they could easily have provided a mutual alibi or organised diversions for the workmen, as and when required. The smuggling of discrete objects into the Lavant Caves would only make sense if *one* of the directors was generating the fraud without out the knowledge of the other. Having an accomplice who was 'in on the fraud' was also potentially hazardous; opening the main perpetrator to exposure or blackmail. That Lewis suspected something was remiss seems clear enough, from the nature of his later relationship with Dawson, but in the course of orchestrating his deceptions, as we shall see, Dawson always worked alone.

The strategy behind the Lavant Caves deception was, therefore, simple enough, although there was a curiously scattergun approach to its implementation: seeding the cave floor with a disparate range of artefact types from a variety of periods. This may be a reflection of Dawson's relative naivety at this stage in his antiquarian career, being only a recent convert to the field of archaeology, or it may have resulted from a wish to make the caves appear *more* significant than just a simple flint mine. True, the finds range as reported appeared to indicate the quarry had begun life as an area of Neolithic flint extraction, but the addition of later 'finds' certainly provided the site with a currency greater than any other known flint mine, suggesting major reuse in the Roman and later Medieval periods.

The chronological depth provided by the placed artefact assemblage made the Lavant Caves appear extremely interesting to an outside audience, certainly more than a mere chalk quarry, which is what, in reality, the caves were. Time, depth and varied use were aspects of the site that Dawson himself was later to emphasise in public lectures and interviews, but a formal publication of the caves was never forthcoming. His reticence in publishing an excavation report, however, can easily be explained, for the Lavant Caves were, from Dawson's own perspective, an important 'discovery', which was much heralded in the pages of the local press and amongst national archaeological societies. All this helped establish him as an antiquarian of some repute, but, ultimately, the caves would not have been a site that he would have wanted to dwell on for long. Publication would certainly have required a full and detailed interpretation of the results, perhaps drawing attention to any inconsistencies, errors or shortcomings in the data set.

The Dungeon

The existence of tunnels or 'dungeons' beneath Hastings Castle had long been known, tours forming a major part of any visit to the medieval fortifications. In 1872 Dawson himself, then aged 8, explored part of the subterranean chambers in the company of a former custodian, whilst a brief description of the tunnel had been made in 1877 by the Reverend E. Marshall of St Leonards-on-Sea in the pages of the *Sussex Archaeological Collections*. So popular had visits to the tunnels become that in 1878 'the door to this excavation was strictly closed upon the public, the custodian becoming tired of taking people over; the atmosphere was bad and the steps then dangerous'.

All this changed in 1894 when Charles Dawson and John Lewis persuaded the then owner, Lord Chichester, to be allowed to reopen the tunnels and clear the major passageway. Quite why the two antiquarians had singled Hastings Castle out for their particular brand of archaeological investigation is unclear, although after the Lavant Caves their expertise in subterranean fieldwork was not in doubt, whilst the castle itself was very close to the Dawson family home in St Leonards-on-Sea.

Having got through the locked doors, Dawson and Lewis were faced with two passages running in a north-east and south easterly direction. The north-east passage descended almost immediately via a series of eight steps, 'hewn out of the sandstone rock, which are worn to such an extent that they can only be descended with great difficulty'. Having constructed a wooden staircase over the badly eroded original, the two followed the tunnel as it veered to the south-east, via a small vestibule or chamber. Along the western edge of this south-facing chamber, the basic outline of a round-headed arched passage had at some date been cut, presumably, Dawson and Lewis speculated, with the intention of connecting with the southern passageway. From the vestibule, an archway led through to another passage running in an east-north-easterly direction.

Passing a small arched recess, which was interpreted as a second abortive attempt to drive a tunnel in a southerly direction, Dawson and Lewis entered what they described as the main gallery. This measured 17ft (5.18m) in length, 4ft (1.22m) in width and was 9ft (2.74m) high. At the north-eastern end of the gallery, a short section of tunnel led first to the north, then finally west, connecting to a 'peculiarly domed chamber' measuring 7ft (2.13m) high. Two small recessed areas 'like fire-places' were noted in the southern wall of this final chamber.

Following the second tunnel from the south-east of the main entrance door, Dawson and Lewis observed that the steps were not as heavily worn as those in the northern passage, a large quantity of debris, presumably derived from roof collapse, having blocked all access. Finds within the debris included both human and animal bone, together with a small fragment of carved white marble. Unfortunately, the extent, quality and final destination of any finds retrieved during the course of the clearance of the Hastings tunnels, remains unknown.

So far, so familiar. The Hastings tunnels had certainly proved to be more dramatic than the chalk-cut caves of Lavant, but, in truth, the 1894 exploration was not adding much detail to the history of Hastings Castle, as Dawson must have realised quite early on. Furthermore, the task of clearance was not proving easy, involving much difficult and back breaking work for Dawson, Lewis and their team of (unnamed) workmen. The record produced by both directors was, however, totally new; John Lewis' inked drawings, which accompanied the later report, added detail to aid the understanding of the subterranean feature whilst Dawson's own photographs, taken 'with the magnesium light', furthermore aided overall interpretation. Once again, however, Dawson had started work on a site which was not proving to be quite what he had hoped for. What the tunnels need was a bit of 'spicing up'.

The 1894 deception set out beneath Hastings Castle was, of necessity, different from that conceived in the Lavant Caves. Here, the subterranean passageways were clearly Medieval, so the inclusion of 'early finds' would appear incongruous whilst the possibility for finding something dramatic was lessened by the fact the main tunnel had been at least partially emptied some years previously. Only one aspect of the tunnels provided Dawson with an outlet for his ambition, and this he seized with both hands.

At the commencement of work, both Dawson and Lewis had appeared critical of earlier interpretations of the tunnels that stressed a dungeon or prison interpretation, but in an area of rock face discolouration on the southern wall of the main gallery in the north-eastern tunnel, Dawson finally allowed his imagination to run wild. This staining, he conjectured in the published report for 1894, 'resembled two shadows of human bodies on the wall, falling side by side between the so-called staple holes'. The sketch that accompanied the article showed the human-shaped markings in some detail, suggesting that they were indeed caused by the placement of bodies against the side wall of the underground chamber. Dawson himself compiled the sketch, the published report clearly noting that 'this discoloration … was certainly plainly visible when Mr. Dawson first saw the cavern, in 1872, especially when a light was held in a particular position near the wall.' By the time of the 1894 clearance, however, the images appear to have faded, Dawson and Lewis commenting that 'strange to say, in our many recent visits to the dungeons, we have never yet again observed this phenomenon, and it now appears entirely lost.'

In other words, Dawson was claiming that although the 'shadow-markings' themselves had vanished by the time that he and Lewis explored the tunnels, they *had* existed and he was now the *only* person who not only remembered them, but could reconstruct their shape and form. Presumably, as no other illustration of the features is known to exist, Dawson's published drawing *must* have been compiled from memory. Dawson was, however, a mere 8 years old when he had seen the shadows for the first (and last) time, this being at least some twenty years *before* he finally published the drawing. He could, of course, have sketched the figures during, or shortly after, 1872, though it must be said that the illustration does not appear to be the work of a young boy. This is an amazing, and rather breathtaking, statement to make, for he was claiming nothing less than a definitive interpretation of an archaeological feature that was not only now destroyed, but which no one else could claim to have seen or could remember having been there in the first place.

This brings us to an interesting conundrum, for the published article appears to play down the significance of the shadow-markings, even going so far as to imply that they may have been crude forgeries (generated by 'applying oily substances to the wall'), whilst the illustration which accompanies it seems to conclusively favour their existence. We could of course infer some form of foul play: perhaps the markings were generated by Dawson and/or Lewis to support the prison hypothesis. This would seem plausible were it not for the fact that the published report so vigorously attacks any suggestion that the tunnels had been used for such a purpose.

It is worth noting here that although some discoloration on the walls of the main gallery had been first observed by the Reverend E. Marshall in 1872, Marshall himself had made no attempt to define or describe the shape of these 'shadows', other than to note their possible origin (from 'exuding' corpses). It is, in fact, Dawson who created the first, and to date *only* depiction of the 'shadows' as representing two discrete and clearly defined humans, apparently held securely in place by metal staples. Of course we have no way of assessing the validity of Dawson's sketch, as it would appear to have been drawn entirely from memory. Perhaps then it was Lewis, who did not see the original staining in 1872 and was unable to see it in 1894, who was the voice of caution in the final published report, even going so far as to suggest that the figures may have been faked. Such an explanation could explain Lewis' scepticism, and the final inconclusive nature of this part of the jointly published report.

The Axe

In 1894, less than two years after joining the Sussex Archaeological Society, Charles Dawson was not only conducting his own fieldwork, but was also publishing the results of antiquarian investigations across the county of Sussex. His conversion to antiquarian archaeology seemed complete and the rapid publication of papers would normally be commended were it not for the fact that the articles themselves appear to have been filled with a generous concoction of truths, half-truths, lies and almost total fabrication, being generated from an almost complete absence of verifiable data.

In 1893, the newly founded Committee of the Hastings Museum Association had begun to acquire collections from various local antiquarian collectors. One such collection comprised an extensive quantity of pre-historic worked flint assembled by Stephen Blackmore, resident of Frost Hill Cottage, East Dean near Eastbourne. Charles Dawson, who inspected the material, judged it to be 'one of the finest collections of Neolithic flint implements in private hands in England'.

In the report published for the 1894 edition of the *Sussex Archaeological Collections*, Dawson noted that, whilst discussing the nature of the collection with Blackmore and terms for which the Hastings Museum would be allowed to purchase elements for their loan collection, his eye was taken by 'a drawing of a haft bearing an implement *in situ*'. As evidence of how prehistoric flint tools were originally hafted was extremely rare, Dawson was intrigued and asked Blackmore about the history of the drawing. Dawson, in transcribing Blackmore's comments, noted that 'some years ago' whilst digging on the cliffs at Mitchdean, East Dean, Blackmore had found a number of flint tools 'namely a flint arrow head, along pick-shaped worked flint, a peculiarly worked flint, much curved, bearing a notch at one end, rather resembling a netting needle, and a long polished stone, probably for honing or polish-ing purposes'. It was close to these that the East Dean resident found a flint implement still 'lying in its wooden haft'. The haft, unfortunately 'crumbled at the touch, and all attempts to save it proved futile'.

The unavailability of the artefact for further study was, of course, unfor-tunate, but luckily Blackmore 'was able to make a drawing of this most interesting discovery', which Dawson was able to modify for the purposes of the publication. It is important to emphasise that the figure accom-panying Dawson's article was *not* the original drawing made by Stephen Blackmore, rather it was a loose *interpretation* of that first illustration. How close to the original drawing, or indeed for that matter the original find from the Mitchdean cliffs, Dawson's version was, we shall unfortunately never know. The same note of caution may also be added with regard to the other flint implements appearing in the figure accompanying Dawson's article, most of which do not appear to have any parallel with the known Neolithic tool types recorded from southern Britain.

Dawson was able to supply a significant amount of detail concerning the hafted tool, following his conversation with Blackmore, noting that 'the implement was received in a horizontal groove and on one side of the shaft near the head. Above it, in the head of the haft, appeared two small stumps, apparently where small shoots had been trimmed off the wood. Below the implement were a number of grooved rings running around the haft. The object of the stumps above the implement, and the grooves beneath, appear to leave no doubt that they were made to receive the cross lashings which secured the implement in its groove in the haft. The blade of the implement itself was inclined slightly downwards, and the haft curved back slightly in the centre'.

Such detailed observations and comments would appear, on the face of it, plausible enough, especially if the haft in question had actually survived as a complete artefact. Unfortunately, as has already been noted, the object had *not* survived at all, having been destroyed at the very moment of its discovery. Dawson's record is, therefore, based entirely upon Blackmore's memory of events conducted 'some years ago'. John Evans raised this point shortly after the publication of the axe head in his book *Ancient Stone Implements*, where he noted that 'neither the description nor the drawings of this … are such to inspire confidence.'

What is more surprising, however, is the recorded *condition* of the hafted flint, given its provenance, within the well-drained chalk soils of East Sussex. Had the soils that Blackmore is supposed to have disturbed been in any way waterlogged, creating the anaerobic conditions that reduce decomposition of organics, then the discovery of such an item would not appear out of the ordinary. Dawson does note that the wooden haft was 'carbonised', suggesting that direct contact with fire had in someway preserved it, but unfortunately fails to elaborate on how such a fire could have left the head and surrounding area strangely unburnt.

The published evidence supplied by Dawson to account for Blackmore's hafted stone axe is not at all convincing. Given the circumstances that he provides for the discovery and eventual reporting of the artefact though an interpretation of a memory, then this is perhaps unsurprising. An objective account of what was really found on the cliffs at Mitchdean is unlikely ever to be forthcoming, and the version that we possess (provided by Dawson) could easily have been distorted, garbled or inflated at any stage in its recounting. It is worth reiterating, however, that we only possess Dawson's description of both the artefact and the events surrounding its discovery. That Stephen Blackmore existed, there is no doubt, but the items credited to him by Dawson (and illustrated for the *Sussex Archaeological Collections*) have never been seen since Dawson described them. It would seem that his imagination and skill at manipulating the truth had scored Dawson his first proper archaeological paper.

The Boat

The second of Dawson's artefact-based articles to appear in the *Sussex Archaeological Collections* for 1894 concerned the discovery of an ancient boat. This particular sea-going vessel had, Dawson related, first been discovered 'about thirty years ago, embedded in the blue Wealden clay on the sea shore' just to the west of Bexhill. The boat had apparently been first exposed in the 1860s, during a particularly violent storm. Following this dramatic natural event, a coastguard walking the stretch of the shore saw the boat together with some bones, including 'a perfect skull of a horse of small size'.

Unfortunately, the boat itself was subsequently covered over by the shifting coastal sand, whilst the unnamed coastguard sold the horse's skull to a gentleman (unknown) from Hastings 'for seven and sixpence'. Dawson observed that the ancient boat had been periodically revealed and submerged until the winter of 1887, when a massive displacement of sand had exposed it almost entirely. At that point the vessel was observed by one Jessie Young, a boat builder from Bexhill who 'immediately took steps to excavate it'. Rather than expose the vessel during the day, Dawson tells us, Young and some friends set about the retrieval 'on a bitterly cold night and with the sea almost at their heels'.

The details and circumstances surrounding the removal of the vessel (date, associated finds, names of individuals involved) though undeniably dramatic are vague and insubstantial. Even the exact position of the find, which could easily have been noted from witness statements, is left unrecorded. Sadly, Dawson observed, 'owing to the rottenness of the wood, the tenacity of the clay, the darkness of the night, and the haste with which the work was necessarily conducted, the boat was very much broken.'

Having extracted the vessel, presumably at no small cost to himself, Young appears to have lost all interest in the project. When Dawson claims to have first seen the boat, it had been left to quietly decay outside Young's workshop. Dawson, recognising the object for what it was, recorded the pieces and, with Young's assistance, attempted an insitu reconstruction, sketching the details for the illustration that accompanied his 1894 article. The final sketch, though undeniably impressive, appears to indicate a well-preserved vessel with none of the decay ascribed to it. This of course implies that Dawson's sketch was more speculative recreation than objective record.

How much of the general description of the Bexhill boat that Dawson provided for his subsequent article is truth, and how much is guesswork or total fabrication, will never be known. Also unfortunate is the delay between the extraction of the boat from the sand and the publication of the article in the *Sussex Archaeological Collections*: possibly as much as seven years. After all this time, should anyone have wanted to re-inspect the vessel to check the validity of Dawson's recreation, they would undoubtedly have found little left to examine.

Dawson remained guarded as to the date and purpose of the small boat. Its basic form, he conceded, appeared to indicate 'a link between the coracle and "burnt out" boat' of popular ancient British tradition, with 'the more modern type depicted in the Bayeux tapestry'. Its importance then was increased as a transitional form, or 'missing-link' in the whole history of British and northern European boat building traditions.

Only one curious element of the whole Bexhill boat story remains. On 21 January 1888, some six years before the appearance of Dawson's article in the *Sussex Archaeological Collections*, the *Southern Weekly News* reported a strangely similar discovery of a wooden boat preserved in the sands of Bexhill. The *News* article, however, identified the finder as Mr Webb, foreman of a construction gang at work nearby. Intrigued as to the nature of the find, Webb instructed his team to remove the object 'with pick-axe and spade', a digging technique which resulted in the boat being extracted in detached pieces. Attempting to later reassemble these rotting fragments of timber proved impossible 'even if the ingenuity of a carpenter would have been equal to the task' observed the reporter for the *News*. The only element of the boat on which it was possible to venture an opinion was the constructional form, which was noted as being 'entirely of oak, the side planks being all pegged together with wood'.

The evidence supplied by Dawson for the Bexhill boat is, as with that surrounding Blackmore's hafted stone axe, really not all that convincing. A lack of any statement regarding the state of preservation of the boat prior to any attempt at reconstruction at once destroys the validity of Dawson's arguments, leaving the reader unsure as to whether Dawson is *inferring* the survival of specific features or accurately recording their presence. The long delay in Dawson's assessment of the vessel and his reporting of it in the pages of the *Sussex Archaeological Collections* would have further clouded the matter as, by 1894, there was probably nothing left of the original boat to inspect.

It is worth reiterating that we only possess Dawson's account of both the boat and the events surrounding its discovery and extraction from the Bexhill sands. That at least one boat did indeed exist, being found in 1888 and reported in the pages of the *Southern Weekly News*, there can be no doubt, but certainty evaporates when we attempt to link that discovery with the one described by Dawson. The circumstances surrounding the finding and retrieval of both boats is surprisingly similar, as is their reported condition following their removal from the beach. Dawson's boat appears to have been found first, in 1887, and had certainly been widely known about since the 1860s, though Dawson's reporting of it post-dated the reporting of Webb's boat by six years. It is possible that the two boats were in fact the same, Young being involved somehow in the original recovery of the vessel from the beach, and later becoming the recipient of the remains once Webb had decided that it was not worth keeping. It is also possible that Young misremembered certain details of the recovery, telling Dawson that it had occurred in 1887, rather than in the third week of 1888. It is possible that he neglected to inform Dawson of Webb's role in the affair. It is far more probable, however, that the story is a complete fabrication.

Dawson appears to have used the tale of the 1888 boat, as found by Webb, as the basis for his article. Carefully he removed details which could implicate him in fraudulent reporting, replacing 'Webb' with 'Young' and keeping the precise details of the discovery vague, but intimating that it had a long history stretching back to the 1860s. The inclusion of an unnamed 'coastguard' and the details of the horse skull (since lost) adds further detail (and authenticity), indicating that Dawson had privileged information. Use of the date '1887' and 'the 1860s' would also have given Dawson's boat precedence over Webb's. Should anyone challenge Dawson's version of events, he could always claim later that he had been supplied with faulty information. He had not, after all, actually claimed that he discovered the boat himself, only that he had been able to interpret it when all others had singularly failed.

Interestingly, as with Dawson's article on Blackmore's hafted stone axe, which accompanied the Bexhill boat in the 1894 edition of the *Sussex Archaeological Collections*, the paper on the boat also deals with an object made of timber which is badly decayed and requires Dawson's expert eye in order to reconstruct its form. In neither case and in neither article was there a real, fully quantifiable object which anyone else could examine and provide an alternative interpretation for.

With his first published articles secure, Dawson felt it was time to be a little more adventurous. Now he needed some genuinely unique artefacts with which to advance his own academic cause.

The Statuette

As Charles Dawson became more adept in the field of archaeological research, his interest in artefact collection, especially ironwork, became more apparent. The 'Dawson Loan Collection', which was ultimately bequeathed to Hastings Museum, contains a variety of diverse iron objects such as a 'hippo-sandal', the tied-on footwear of the Roman packhorse, a variety of nails, and a mass of material from Hastings Castle including 'keys, knives and spears'. Much of this material was used by Dawson to fill the many displays and exhibitions organised across Sussex on ironwork and metalworking in south-eastern England. Only a few of these were ever individually reported upon by Dawson for, rather than publish selected items in archaeological or antiquarian journals, Dawson seemed content to use discrete pieces to show to, and discuss with, various important figures in the antiquarian world.

Throughout the later 1890s, a key element in Dawson's *modus operandi* was developing: his habitual vagueness. Strange that, as a solicitor, Dawson's attention to detail with regard to the specifics of artefact discovery, recovery and location was always frustratingly opaque. In retrospect, of course, it is possible to see this is exactly how he managed to cover up any inconsistencies in the story of how a particular object was found, by whom and when. His earlier fieldwork at both the Lavant Caves and Hastings Castle had contained details of discovery that were fuzzy at best (including unnamed shepherds and unidentified former custodians) but in developing tales of archaeological provenance for his increasingly intriguing finds, Dawson seems to have gone out of his way to deliberately obscure detail. 'Workmen' are frequently left unnamed and precise locations left unclear.

Added to the general vagueness surrounding the initial recovery of particular items, Dawson often introduced a significant time lapse between the finding of an artefact (either by himself or another) and its eventual reporting. Sometimes this lapse could be as long as ten to sixteen years, with Dawson never explaining (nor being adequately interrogated by another) precisely why it took him so long to bring a particular object to the attention of the academic community. A major time lag between retrieval and reporting would, of course, prove extremely useful if one wanted to cloud specific detail concerning provenance and recovery. Sixteen years after the event, it would be likely that few involved in the 'discovery' of a particular item would be able to *precisely* recall how and when it first came to light. Furthermore, even if details of named individuals were provided, and even if it could be shown that they *did* exist (and there was frequently never any independent verification that this was so), it would have been extremely difficult for interested parties to track them down. An added bonus would be that the slag heap / gravel pit / road working / hedge planting / coastal sand from which an object in question was supposed to have derived would, by the time the report was made public a decade later, probably have changed out of all recognition or have been completely obliterated, making re-excavation or re-examination of the surrounding area impossible.

As a solicitor, Dawson certainly knew how to bend the law in multiple ways. When it came to bending the truth, mis speaking, obfuscation, wilful ambiguity and plausible deniability, he was truly the master.

Dawson's first academic target was both important and well placed. Early in 1893, Augustus Wollaston Franks, Keeper of British and Medieval Antiquities at the British Museum and President of the Society of Antiquaries, was presented with an unusual artefact: a small and rather corroded iron statuette. The statuette was clearly Roman in style, and appeared to represent a miniature copy of a horseman from the Quirinal in Rome. The owner of the piece, Charles Dawson, confirmed to Franks that the object had found beneath a slag heap from the Roman ironworking site of Beauport Park in East Sussex by an unnamed workman digging for material to surface a local road. Dawson's evident excitement concerning the statuette was that, in his view, it was made of cast iron and was not of the more usual wrought (or hammered) form.

In the late years of the nineteenth century, it was believed that highly carbonised, or cast iron had not been produced in Europe before the end of the fourteenth century, and had certainly not appeared in the British Isles until at least the end of the fifteenth. Such iron was used primarily for making castings, but in its finished form was quite hard and brittle and could, therefore, not be welded or forged. In order to produce cast iron, by melting the iron in iron ore, a metalworker would have to generate a temperature of around 1200 degrees centigrade; something which was considered impossible before the advent of blast furnaces. Before the blast furnace, iron was smelted in 'bloomery hearths' (usually attaining temperatures of just over 900 degrees centigrade), which reduced the material to mass of sponge iron. 'Bloom' iron, containing less than 1 per cent carbon, is tough and malleable and can easily be worked in a blacksmith's forge, where it is referred to as 'wrought iron'. Given these facts, any iron worked in the Roman period *ought* to be of the wrought variety.

James Rock, an archaeological fieldworker who observed the destruction of the Roman ironworking site at Beauport Park in the 1870s, certainly believed that the Romano British ironworkers had the capability to generate cast iron, and said as much in his article published within the 1879 edition of the *Sussex Archaeological Collections*. If *molten iron* was being found on Sussex iron-working sites from the Roman period, it would only mean that the ironworkers of this period had indeed attained temperatures in excess of 1200 degrees centigrade and had been able to liquefy iron for use in moulds. If Dawson's Roman statuette had indeed been cast, then the whole history of iron-working in Europe would be turned on its head.

We now know that Dawson's evident excitement concerning the Beauport statuette was due to the fact that he *knew* it was indeed made of cast iron. Unfortunately, he also knew that the piece was not a genuine Roman artefact, it being a modern copy of an ancient original. It is doubtful whether Dawson manufactured the statuette himself, that would require a great degree of metallurgical experience (and good access to a furnace), but it is likely that he was able to successfully doctor a modern item in order to enhance its plausibility; to make it appear a genuine Romano British antiquity. This was, he would try and convince people, an internationally significant 'transitional' find: the first example of modern smelting techniques. It was, of course, a huge gamble. If he had been found out, and the piece shown to be a modern forgery, all of his other 'finds' and 'discoveries' would naturally come under both suspicion and careful scrutiny. Dawson, though, was able to employ his standard tactic: obscuring the exact nature of both find and provenance. If the artefact was shown at any stage to be a forgery, he had already created a story that would allow him to plausibly deny any part in the overall falsehood.

In order to officially determine whether the Beauport statuette had been cast in a mould (as Dawson was suggesting), or whether it had been hammered into shape (as one would expect for a Roman antiquity), the specimen was submitted to W.C. Roberts-Austen, professor of metallurgy at the Royal School of Mines and assayer to the Royal Mint. A small sample of the figure (6.77 grains) was taken and dissolved by Roberts-Austen, who observed that, because of the minute quantity of carbon recorded, he had 'no hesitation in saying that the figure was not made of cast iron, but was of wrought, malleable iron'. Dawson, naturally, disagreed.

When, on 18 May 1893, the statuette was exhibited at a London meeting of the Society of Antiquaries, it was described as being made of wrought iron, presumably in the light of Roberts-Austen's analysis. Dawson attended the meeting with his object, but could not speak in its defence, as he had yet to be elected a fellow of the Society (something that he was already at pains to rectify). Presumably, however, Dawson supplied Charles Hercules Read, Deputy Keeper of British and Medieval Antiquities at the British Museum, who introduced the artefact to the meeting, with all the necessary details, such as they were, concerning its original 'discovery' and perceived antiquity. Read stated that the piece had been found by a still unnamed labourer, employed to quarry material from the Roman site, 'at a depth of 27 feet' into a slag heap. The work of Roberts-Austen implied that the statuette had been manufactured by hammering and chiselling, such a slow process apparently making it 'inconceivable that it can have been done at a time when the casting of iron was practised'. To Read, this observation appeared to confirm that the artefact was Roman date, rather than having been made at any later time.

The questions and comments that followed the presentation were not overly enthusiastic. A.H. Smith displayed a small bronze statuette to the assembled meeting, which was of the same basic design as Dawson's exhibit. This figure of a horseman had, Smith noted, only recently been purchased in Orange, southern France, in the belief that is was a genuine Roman artefact. Following a close inspection of the figure however, Smith was of the opinion that the Beauport statuette was probably a modern replica, possibly manufactured as a souvenir in Rome. A.S. Murray, Keeper of the Department of Greek and Roman Antiquities at the British Museum, noted that the fine modelling evident in Dawson's iron statuette, combined with the observation that it appeared to be a reproduction of a genuine Roman piece, were strong arguments *against* its authenticity. Sir John Evans noted the great similarities between Dawson's iron figure and the 'modern' replica belonging to Smith, adding that 'suspicion might be aroused as to their belonging to the same category'.

Sadly, Dawson's reaction to this overtly negative response is not recorded, though four months later, on 13 September 1893, he wrote to W.V. Crake, then secretary of the Hastings Museum Association, that 'they rather disputed my iron-statuette at the Society of Antiquaries so the put me on my mettle a bit'. Also contained in the letter to Crake is a more explicit reference to the circumstances surrounding the initial discovery of the iron figure, which had not previously been circulated. Presumably, since the meeting in London, Dawson had been thinking about the voices of doubt and realised that, to achieve a more favourable response to his 'Roman' figure, he would need a better version of the story surrounding its recovery. 'On Sunday' the solicitor wrote, 'I managed to collect evidence at Westfield about my little cast iron statuette. I found the labourer who dug it up and who says it was with coins of Hadrian's time. The coins found their way into the collection of Mr Rock – late Rock and Hawkins – White Rock. Do you think you could discover him? I believe he had many losses after retiring. I should be very much indebted if you could – and perhaps he has other local specimens he would cede to the Hastings Museum.'

This letter is interesting for a number of reasons. First, it suggests that a key concern of the Antiquaries in London had indeed arisen from the specific lack of detail surrounding the finder. The details provided to Read had been vague, possibly fuelling the suspicion that the piece was fraudulent. If Dawson could obtain a sworn testimony as to the circumstances of the discovery, as well a more precise location as to the spoil heap from which it had been 'buried', then there was a good chance the piece may finally be accepted as a genuine Roman artefact. Secondly, the comment that the statuette had been found 'with coins of Hadrian's time', was something that could be directly linked with an earlier, authenticated discovery made by James Rock in the 1870s. In an article entitled 'Ancient Cinder-Heaps in East Sussex', published in the 1879 edition of the *Sussex Archaeological Collections*, Rock had observed: 'I have in my possession two coins of bronze, which were also found among the cinders – one of Trajan, the other of Hadrian. Both are in good preservation, especially the later. These would seem to fix the date of the cinder-heap at a somewhat early period of the Roman occupation.'

Rock goes on, in his description of artefacts recovered from Beauport Park, to mention a bronze ring, a bronze spoon found 'at the bottom of the cinder-heap' and numerous fragments of pottery, including fine red Samian ware. Quite why the statuette, surely the most prized find from Beauport Park, was not mentioned by him is unclear. Perhaps, gambled Dawson, people may think that Rock was unaware of its existence, the workman who found it simply not passing it on to the collector. Perhaps the coins and figurine, though found in the same spoil heap, were discovered at different times or by completely different labourers (Dawson's account does not specifically state that it was *the same* workman that made both discoveries). Had the coins and statuette been found at broadly the same time (in the 1870s), however, it would strange that the workman or men in question had not attempted to sell the piece (as most were paid extra for any interesting discoveries). Why did the, as yet still unnamed workman, hold on to the artefact until the 1880s before selling it to Dawson? This was an anomaly in Dawson's story that was to remain unexplained.

Also of note in Dawson's letter of 13 September 1893 to Crake is that the solicitor remained convinced, despite the analysis of Roberts-Austen, that the Beauport Park figurine was in reality made from cast iron, not wrought as had been suggested. One last piece of evidence was, Dawson commented, now available to him which could change the accepted view of the Society of Antiquaries: 'Fortunately and almost wonderfully, an iron spearhead cast in a mould as in the Bronze Age, has been found in an ancient iron pit in the west of England and is now on its way to me. The question is had the ancients the knowledge of *casting* iron. I hope it will be a knockdown blow.'

Nothing more was ever heard of this mysterious iron spearhead. Either Dawson never received it, or it proved not to be the 'knockdown blow' that he had implied it would be. More probably it never existed other than in Dawson's fertile imagination.

Given its rather less than impressive impact, the Beauport Park statuette was retired by Dawson until the 1903 exhibition of Sussex ironwork, which he was to organise in Lewes on behalf of the Sussex Archaeological Society. In the report that followed the exhibition, published in the *Sussex Archaeological Collections*, Dawson provided the most detailed account to date of the discovery of the iron figurine. Now, after the space of a further ten years, he was able to state that: 'the specimen was found by one of the workmen employed in digging the iron slag for road-metal about the year 1877. His name is William Merritt.' Adding, 'he lives at King Street, Seddlescombe Road, Westfield.' 'All the workmen engaged in digging were in the habit of picking up any of the more important specimens,' he went on, 'and keeping them for certain people who were interested in the

discoveries at the time. The work extended over many years, and the prin-
cipal slag heaps were disposed of. The author, who had been recommended
in the year 1883 to see Mr Merritt about some geological specimens, procured from him, with other specimens, a small, much corroded statuette,
all of which he stated that he had dug up in the slag heaps of Beauport …
The author, as far as possible, took considerable trouble to settle the *bona
fides* of the discovery, and received from Mr Merritt a written account
authenticating it.'

It is implied, although never stated, that the 'written account' authenticating the find was obtained from the 1893 meeting with the Westfield
labourer, which the solicitor refers to in his letter to Crake. Why he had
waited until 1903 to 'go public' with these details, especially as they seemed
so important to him ten years before, is also left unclear (and, fortunately
for Dawson, unquestioned). Although, for the purposes of introducing yet
another time lapse, they seem inherently clear. Sadly, but perhaps unsurprisingly, there is no record of Merritt's written account. We have only
Dawson's word that it (or indeed Merritt himself) ever existed.

Dawson went on in his 1903 report to describe the Beauport Park
statuette in detail, noting that it represented one of the finest (and most
ancient) of materials in his exhibition of Sussex ironwork. He related the
fallout of the 1893 Society of Antiquaries meeting, adding that 'it was
afterwards examined by several different experts with great diversity of
opinion, some stating that it could not be Roman, because the Romans
had no tools capable of producing it in wrought-iron, others dismissing
the matter by stating that if it was of cast-iron it could not be Roman'.
Introducing an element of apparent objectivity to the *Sussex Archaeological
Collections* article, Dawson observed that:

> wishing to decide the question definitely the author sent the statuette to Dr
> Kelner, of the Royal Arsenal, Woolwich, who has, of course, great experience
> in the analysis of iron, for his determination on analysis. A portion of the
> metal was removed from the interior of one of the leg stumps. The Arsenal
> workman who bored it stated that it cut like cast-iron. Dr Kelner reported
> that there was not the slightest doubt as to its being of *cast*-iron. Under these
> circumstances, and in the absence of further evidence, the author is disposed
> to claim that this little statuette is Roman, or Anglo-Roman, and the earliest
> known example of cast-iron in Europe at least.

Thus Dawson was finally able to proclaim that he had indeed located the earliest example of cast iron in Europe, proving that, not only did the Roman state possess far greater technological capabilities than had previously been supposed, but also that the 'missing link in the history of the iron industry', between primitive methods of hammering and the more advanced use of blast furnaces, had been found. Unfortunately, the impact of this announcement was somewhat lessened by the fact that it was published in a county-based journal (rather than one of more international standing), a whole decade after Dawson had first produced the artefact. Of course the confirmation that the statuette was indeed of cast iron (as Dawson had of course known all along) was one thing, but Dr Kelner's report had not said that the statuette was *definitively* Roman. That was a leap of logic that only Dawson was prepared to make.

Writing in the early 1930s, Ernest Straker, in his book *Wealden Iron*, the definitive study of the iron industry of south-eastern England, commented that: 'Notwithstanding Mr Dawson's belief in the authenticity of this find, there are some doubts on the matter. The sale of the objects found was a valuable source of income to the diggers, and it is possible that deception may have been practised. From the context it is evident that similar bronze figures have been produced, and a replica in modern cast iron would not be difficult to cast and corrode by burial.' Straker did not offer an opinion as to who could have conducted the deception, as at that stage there was no widespread doubt as to the wider activities of the then deceased Charles Dawson, but the implication was that it was either Dawson or the finder, the mysterious Mr Merritt.

In the early 1950s, following the revelation of the Piltdown hoax, Robert Downes of Birmingham University took a thin slice was from the damaged right leg of the statue. Microscopic examination indicated that, just as Dawson had noted in 1903, the material was cast iron, but the relatively high level of sulphur present indicated that the statuette had been produced in a coke-burning furnace. 'If this were the case,' Downes observed, 'the statuette could not have been made before the eighteenth century and not in Sussex at any date.' The nature of surface damage to the piece, which had been used as proof of its antiquity, was, it was suggested, potentially indicative of 'burial for many hundreds of years', but, rather ominously, the pattern of corrosion was 'equally consistent with burial for one year at a suitable site'. Sixty years after it's initial reporting, the Beauport Park statuette was finally revealed to have been a hoax.

Although Dawson's iron figure had not generated quite the sort of scientific impact that he had hoped, it had at least attracted the attention, and brought the Sussex solicitor within the orbit, of Augustus Wollaston Franks and Charles Hercules Read, who, apart from their positions in the British Museum, were respectively the President and Secretary of the Society of Antiquaries. Dawson had even had his prized exhibit presented before a meeting of the Antiquaries in London, quite an honour for an amateur investigator.

The Horseshoe

In 1903 Dawson supplied a unique form of ancient horseshoe for the Lewes exhibition on Sussex ironwork. The object was illustrated and briefly discussed by the solicitor within his report on the exhibition published in the *Sussex Archaeological Collections* for 1903. Here, the artefact was described as being: 'a somewhat heavy slipper-form, the plate slightly moulded to the frog of the foot of the horse, and the front edge perforated with nail holes somewhat in the manner of a modern shoe. The back centre of the plate was flanged upward, the flange terminating in a hook-shaped piece as if used to strap the hinder part of the shoe to the horse's hoof.'

The object appeared to represent a transitional or intermediary stage in horse footwear, between the slipper or hippo-sandal (a Roman temporary shoe often tied directly onto the hoof of an unshod horse) and the more usual nailed variety. The sketches provided by Dawson for the purposes of the publication are unfortunately lacking in any kind of detail, making an objective assessment of the piece rather difficult. The exact context of discovery is, furthermore, never established in detail, though Dawson observed that it had originally been found by person or persons unnamed 'associated with the piles of an ancient bridge … which had been superseded by another wooden bridge time out of mind.' As with a number of Dawson's 'finds', it is the deliberate blurring of detail, especially with regard to both finder and find spot, that ultimately makes any interpretation difficult. Dawson skilfully infers association with 'the piles of an ancient bridge', implying, but never explicitly confirming, a Roman origin for the artefact.

The find was presented to Sir Wollaston Franks of the British Museum for identification at some date before 1896 (Franks died the following year making confirmation of his meeting with the solicitor rather difficult to establish), Dawson later reporting that the scholar 'was disposed to regard it as a development of the type of the 'Roman shoe' or 'hippo-sandal'. Robert Downes, who examined the evidence for the Uckfield horseshoe in the mid-1950s, was extremely sceptical, noting that 'even in the existing state of knowledge, such a 'missing link' must have appeared superfluous' a front nailed, rear tied shoe combining 'mutually contradictory principles as it flapped about' the horses hooves.

The artefact itself has, unfortunately, since been confined to oblivion. Holland's article notes that, in 1896, the shoe was in the British Museum, presumably a temporary affair whilst it awaited identification. Its last appearance was as part of Dawson's 1903 exhibition at Lewes, after which it finally disappeared from view, probably because, as with the Beauport statuette, it no longer served any useful purpose. No further examples of such horseshoes have, perhaps unsurprisingly, come to light since 1903.

The Spur

In June 1908, Charles Dawson brought three iron objects to a meeting of the Society of Antiquaries in London. Two of the three were described by him as examples of 'prick spur' found at Hastings Castle whilst the third was simply categorised as 'an iron object from Lewes Castle'. Dawson never supplied any information as to the circumstances under which these pieces were recovered, though we may presume that, if genuine, the Hastings Castle artefacts were derived from one of the many investigations conducted there by Dawson and Lewis prior to 1900. The object from Lewes Castle is less easy to provenance, though given the date of the London exhibition it is possible that it had been found in the back garden of Castle Lodge, the Lewes town house that Dawson and his family had moved into during the spring of 1907.

The key point of interest surrounding the Lewes Castle object is its identification and interpretation. When Dawson exhibited the piece at the meeting of the Society of Antiquaries, it is clear that only the accompanying objects from Hastings Castle had by then been securely identified as spurs, being similar to those worn by Norman knights in the Bayeux Tapestry. When the object from Lewes was displayed by Hastings Museum in 1909, it too appears to have been identified as a prick spur.

Spurs, designed primarily to control a horse thorough the digging of something sharp into its flanks, made their first widespread appearance in Britain at the time of the Norman Conquest, though certain Roman cavalry units may have made use of them earlier. The first recognisable form of spur is the prick spur: a simple immovable point fastened to the heels of the rider. Over time the form of the spur developed and evolved, whilst the object itself became a sign of prestige, not only in the ownership of horses (with all its implied status), but also as a sign of chivalry, later references to becoming a knight being described as 'winning one's spurs'.

The Lewes Castle prick spur is made of wrought iron and possesses a rather lethal-looking sharpened point where it, presumably, made contact with the horse. In 1954, Rupert Bruce-Mitford, then Keeper of the Department of Romano British and Medieval Antiquities at the British Museum, observed that 'the object is certainly not a prick spur and is nothing we can recognise'. In fact the artefact does look rather overly dangerous and, if employed as a spur, would undoubtedly have badly injured the flanks of any horse to which it was applied. Whatever the origins of the piece, it would not appear to be Medieval although Dawson, not unexpectedly, made much of the close spatial association between the (vague) provenance and the former location of the Lewes Castle's tilting ground (a space now occupied by the bowls club).

The Hammer

Charles Dawson's collection of archaeological discoveries was not restricted solely to iron artefacts. During his lengthy career as amateur antiquarian he collected a variety of curious objects made from bronze, antler and bone. Some of this material he collected himself during the course of field examination, other pieces he procured from antiquarian collectors, labourers, antique shops, pawn brokers and workmen.

In 1905, George Clinch, writing in the 'Early Man' section of the *Victoria County History of Sussex* (Volume 1), recorded a number of 'curious objects of deer-horn' preserved in the Dawson loan collection of the Brassey Institute in Hastings. One piece particularly caught his attention, comprising a section of antler 'pierced in the middle by a nearly square hole'. Although Dawson himself never appears to have commented publically on the artefact, he supplied some information to Clinch who was able to report that 'it is said to have been found in the submarine forest at Bulverhythe, halfway between St Leonards and Bexhill'.

Dawson had, for the purposes of the Brassey Institute accessions register, identified the artefact as a hammer, though, as Clinch observed, as a hammer 'it does not seem to be particularly fitted'. This is clear enough; when examining the artefact, it is apparent that, although the central squared hole could easily have accommodated a haft or handle, the absence of a flattened or battered head (or indeed of any obvious point of impact) would seem to rule out Dawson's interpretation as a hammer or instrument of percussion. Clinch could find no real parallel for the piece, commenting that it may have been used as an equestrian cheek piece (or a form of toggle connecting reins to a rope or sinew mouthpiece) or bridle bit.

As the object was both unstratified and ultimately unassociated with any other datable material other than the submerged forest, a landscape of oak, alder and hazel tree stumps drowned when the sea levels rose towards the end of the Mesolithic some 6,000 years ago, Clinch was unable to assign the object to a specific time period. As a consequence, the curious item appeared under the heading of 'miscellaneous antiquities', though assumption was that it was probably either of Neolithic or Bronze Age date.

Both Clinch and Dawson favoured a prehistoric date for the item, based primarily on the material used (red deer antler was a favoured source during the Mesolithic, Neolithic and Bronze Age) but also as it had been found within the remains of the Mesolithic submerged forest (though the details of the association were not, as with the Beauport Park statuette and the Uckfield horseshoe, totally clear). Unfortunately, neither Clinch nor any other antiquarian working in Sussex at the time, appeared to have considered the nature of the central 'hafting' hole which, on close inspection, holds the key as to the date of manufacture.

Two aspects of the cut require obvious attention: first, the hole is both rectangular and straight edged, suggesting that the bone was already quite old (perhaps almost fossilised) when worked and not a freshly discarded piece; secondly, the edges of the cut itself suggest that it was made with a steel chisel, the imprint of which is still visible in the corners of the squared hole. In other words, the antler itself may be ancient, but the rectangular cut through its central section could not have been made before the eighteenth century AD.

If we take Dawson at his word, then the object was retrieved, presumably at low tide, from amidst the fossilised tree stumps of a submerged forest.

Once again, however, Dawson supplies no extra detail as the circumstances of discovery, nor of whether the object was lying on the surface when discovered or (at least partially) embedded within sediments. It is clear, from the surface condition of the piece, that it had not been exposed to a marine environment for any length of time; the edges of the cut hole are too sharp and there is no evidence of salt-water damage or tidal battering. It is likely, therefore, to have been newly perforated when Dawson found it and his statement that it was found on the sea shore is manifestly false.

The Hoard

Recorded alongside the Bulverhythe hammer in the pages the *Victoria County History* of Sussex were a number of Bronze Age axes or palstaves both whole and fragmentary. No specific details as to the circumstances of their discovery or coherency as a single group are provided by George Clinch, author of the *Victoria County History* chapter on 'Early Man', other than to note that they were derived from St Leonards-on-Sea and 'are now in the possession of Mr Charles Dawson F.S.A.'.

Palstaves represent one of the most common and recognisable artefacts of the British Later Bronze Age (conventionally 1400–600 BC). The term 'palstave' itself is usually applied by modern archaeologists to a form of bronze axe head with a flared, curving edge and distinct shoulders. The metal head often possessed a stop-ridge to hold a wooden axe haft in place and occasionally this was supplemented by a metal loop so the blade could be tied directly to the haft with the help of leather strips or twine. Palstaves were gradually superseded in the Bronze Age by socketed axes, a hollow axe head which allowed the direct insertion of the haft.

Both socketed axes and palstaves commonly occur archaeologically in discrete groups or hoards. A hoard is a collection of related metal objects that were buried together at one particular time. Quite why the original depositor of the hoard never returned to claim the material is a question that archaeology itself cannot answer, though a number of possible interpretations are often cited. Perhaps the metal within a discrete group or hoard represented scrap collected by an itinerant smith or metal worker (founder's hoard); perhaps they were quality items kept for later exchange or sale (merchant's hoard); possibly they were acquired as the spoils of war (loot hoard); or represent a range of private effects hidden at a time of civil unrest (personal hoard). They may even represent a form of ritual or religious offering which accompanied the remains of the dead (votive hoard).

Two objects from the tentatively identified 'St Leonards on Sea hoard' were illustrated in the *Victoria County History*: a corroded bronze palstave, slightly chipped along its cutting edge, and an unusual 'bronze socketed object' in slightly better condition. Clinch describes the socketed object as 'evidently only a part of a larger implement and ending in a reversed shield'. Unfortunately neither artefact is discussed in any detail, though the caption that accompanies the socketed object notes a slightly more specific provenance than the palstaves, namely 'the marina' at St Leonards-on-Sea. This, together with the differing states of artefact preservation, casts significant doubt as to whether the two pieces were originally found at the same time or whether they formed part of the same group or hoard.

Unlike the Bulverhythe hammer mentioned above, in the case of the St Leonards hoard, Dawson does provide some degree of provenance. In his 1909 book, the *History of Hastings Castle*, he notes that the material came from 'a fall of the cliff behind the houses at West Marina, St. Leonards on Sea, which took place about the year 1869'. The full extent, nature and original context of the hoard is sadly unknown (if not unknowable). As ,indeed, are the circumstances of recovery, the name of the finder and, more importantly perhaps, the exact association of the pieces credited to it (Dawson refers only the palstaves and the 'runner', probably a small anvil). The differential surface form of the standard mount, when compared to the more heavily corroded palstaves, argue against the mount being part of the hoard. Given that the original discovery of the hoard had been made some forty years earlier, and that Dawson is decidedly non-specific as to where in the 'Marina' the material had originally been located, it is possible he realised that, even had someone wished to check the validity of his statements, this would not be possible. We have, in any case, only got Dawson's word that the hoard ever truly existed.

The Vase

In or around 1886, Charles Dawson presented the British Museum with a rather curious find: a Chinese bronze bowl which he claimed to have found at Dover, in Kent. Once again, the precise circumstances of the discovery were never clarified, though both the artefact and its unusual context were mirrored by a slightly earlier presentation (in 1885) of a Chinese *hu* or bronze ritual vessel, which was given to the same museum by Henry Willett. Willett claimed to have discovered his vessel 'in the Dane John [donjon] at Canterbury', an area where there are no recorded archaeological investigations in the later decades of the nineteenth century.

Henry Willett was a wealthy Brighton-based collector of fossils, rocks, paintings, porcelain and general curios, and a keen advocate of public involvement and awareness. He is credited as being the main impetus in the creation of Brighton Museum and Art Gallery (between 1901 and 1903), following substantial donations from his own private collections, to the people and council of the town. Many of his ceramics and paintings remain on display in Brighton Museum whilst the bulk of his fossils, collection was given to Brighton's Booth Museum of Natural History. The exact nature of his link to Dawson is unknown, although both men moved in similar circles (both having been instrumental in the establishment of important Sussex museums) and clearly knew each other.

Certain doubts surrounding the authenticity of Willett's bronze arise when the various time periods are considered, for the vessel dates from the fourth or third century BC, whilst the influx of Chinese vessels into the northern European market only occurred after the seventeenth century AD when trade with China drastically increased. The date of Willett's *hu* would, therefore, appear to be at odds with its alleged context, making it somewhat unlikely (though not altogether impossible) that the artefact was found within a medieval feature inside a Kentish town. It is, of course, possible that the artefact was brought from China by an antiquarian collector, though how it then materialised in Canterbury is unclear (unless of course either Willett was lying or had been fooled as to its exact provenance). Dawson's vessel, in this respect, appears more suspect, for the solicitor provides even less contextual detail, other than the rather vague comment that it had originated 'from Dover'.

Dawson's bowl measures 33cm (13in) in diameter, being decorated in the style of the Former (Western) Han Dynasty, which lasted from 206 BC to AD 25. A circular panel in the centre of the dish contains two fish set either side of an inscription, which can be translated as 'May you have sons and grandsons'. Robert Downes, a metallurgist by training, and William Watson, then Assistant Keeper of the Department of Oriental Antiquities at the British Museum, examined the bowl in the mid-1950s. Both were intrigued by the surface patination of the artefact: 'a dull olive green ... with a rich incrustation … resembling malachite.' Watson observed that such patination looked odd for a vessel of its date, noting in addition that 'the metal has been beaten, which is not the technique usually found in Han dynasty metal vessels'.

It is apparent from both the surface patination and the process of manufacture, that Dawson's bowl had been created within the last few hundred years, rather than the previous two thousand. This is not perhaps unusual, for copying or mimicking early forms of metal and ceramic artefacts, textiles and texts for economic gain has a long and established history in China, certain 'archaic' bronze forms having been forged at least from the eleventh century AD. Although not a clear 'forgery', in the sense of having been created specifically to dupe the scientific community, one must ask what a post-sixteenth-century Chinese bronze is doing in late nineteenth-century Dover?

Given his interests as a collector and his involvement in public projects (such as the creation of a Brighton Emigration Society to help people move to Canada), Willett seems an unlikely suspect for archaeological fraud. Unlike Dawson, he never aspired to be a great antiquarian and never seems to have specifically craved academic acknowledgement. Furthermore, given what we can deduce from the case of the Pevensey bricks (see below) it seems highly likely that Dawson was the prime mover in this particular curious case of unusual artefact discovery.

Dawson was, it appeared, to be in receipt of a spectacular artefact (in this case a bronze vase from China) which he hoped to show came from a highly unusual source (here Medieval Kent). To do this successfully he would need to create an atmosphere where no suspicion would arise concerning his 'find'. As with his other 'discoveries', Dawson required a dupe, someone of standing with no real or obvious link to himself, who could not only provide a degree of credibility for his find, but who could also conveniently make a similar find first, the earlier find (Willet's) helping to authenticate the latter (Dawson's): a sort of inverse camouflage. A second, broadly similar Chinese vase (complete with dubious provenance) was therefore passed to Willett (possibly via an intermediary) safe in the knowledge that, as a trusted and well-respected collector, such a donation to the British Museum would not be questioned. All suspicion defused, Dawson could then present his piece. In such a scenario, Willett would be no more guilty of forgery than, say, Lewis had been at Lavant or Hastings Castle: he was, for Dawson's purposes, merely the right man at the right time to unknowingly complete a particular piece of targeted deception.

The House

In 1895, in recognition of his hard work and series of amazing discoveries at Lavant and Hastings, combined with his ever-increasing collection of archaeological curiosities, Charles Dawson was elected a Fellow of the Society of Antiquaries, London. By now he was feeling that perhaps he no longer required the help and support of the Sussex Archaeological Society, most of his new 'discoveries' being sent directly to either the British Museum or Hastings Museum and Art Gallery. The society had undoubtedly been extremely useful for Dawson, providing him with a platform on which to showcase his early work, but Dawson was now moving beyond being a regional expert: he hoped to make a larger impact upon the world beyond Sussex. What he did need however, at this point in his career, was an elegant town house which would best reflect his status as premier antiquarian of his generation. It was this desire for improvement in his domestic situation that would ultimately drive both the solicitor and the Society on a collision course.

Since 1885, the Sussex Archaeological Society had occupied Castle Lodge, a grand house situated at the foot of Lewes Castle. The house was leased from the Marquess of Abergavenny, with an understanding that, if the decision was ever made to put the property on the market, the Society would have the first option of purchasing it. When notice to quit the property arrived on midsummer's day 1904, therefore, it was a bolt from the blue. The house was no longer in the hands of the marquess, it had been sold and the new owner wanted the Society and all its possessions out. The council of the Society alerted their members via an emergency notice in the annual journal. Castle Lodge, home to the main offices and museum of the county-based organisation, 'had been sold to Mr. Dawson, and a notice to quit at midsummer 1904 had been served by him on the Secretary'. The council's tone was incredulous: 'This purchase by one of our own members,' they went on, 'and its consequences, took the Council completely by surprise'.

The details surrounding the sale of the property remain unclear, and unfortunately we possess only a one-sided version of the affair, that of the Sussex Archaeological Society, who were understandably outraged to be losing their head office and museum. What made matters far worse in their eyes, was that the eviction was taking place as a direct consequence of one of their own members: Mr Charles Dawson, who, after all they had done for him, was now turning them out onto the street as a landlord would a rogue tenant. Louis Salzman, later president of the Society, described how in his view the purchase of Castle Lodge had not been wholly above board. The central thrust to his accusation was that Dawson had misused both his official position as a solicitor and his connections to the Sussex Archaeological Society to imply that he was buying the property from the Marquess of Abergavenny *on their behalf*. Ernest Clarke, apparently a friend of the Dawson family, later recalled that the vendors, too, were taken aback, for they had not realised until the very last stage of the sale that Dawson was acting in his own interests, using both his official position as a solicitor and his connection to the Society (aided by his unauthorised use of Sussex Archaeological Society headed notepaper) to allow them to assume that he was in some way acting in an official capacity.

We do not, of course, possess Dawson's side of the story here, only the recollections of those who may not have been well disposed towards him and his family. We do not know whether he bought the property in good faith, assuming that the Sussex Archaeological Society were looking for new premises anyway – although it is difficult to see how he could have believed that the move would not have greatly inconvenienced them. His notice to the society to quit the property immediately, removing every last vestige of their occupancy, although entirely within the law, as Dawson would well have known, could have been handled better. The relative secrecy under which the sale had been conducted, combined with the suddenness of the eviction, was hardly likely to win Dawson any supporters within Lewes, the whole affair apparently leaving him ostracised within the antiquarian circles of the county. That, evidently, did not matter much to Dawson, who had calculated that a fine town house was far more important than the social niceties of maintaining a good relationship with local societies.

If Dawson had played on his connections to the Sussex Archaeological Society (as secretary for Uckfield) in meetings with the vendors, increased by his unauthorised use of Sussex Archaeological Societyheaded notepaper in his private correspondence with the Marquess of Abergavenny, this would be a serious matter; a gross breach of professional conduct. As it stands, however, no charges were ever made against the solicitor and so the accusation of misconduct and corruption of office remained unfounded, although the rumours against Dawson continued to circulate in Lewes and beyond.

Whatever the truth of the sale and subsequent eviction, Castle Lodge was eventually cleared and, in the spring of 1907, the Dawsons moved

unopposed into their new home. Over the following years, the architectural features of Castle Lodge became increasingly elaborate, with Dawson
adding external architectural detail, whilst also converting an abandoned
wine cellar into a mock medieval dungeon complete with manacles and
a stone bed. Little is known about the interior of Castle Lodge following
the renovation, though Marie-Joseph Pierre Teilhard de Chardin, a visitor to the property in May 1912, observed that Dawson's stepson Francis
Postlethwaite, then a captain 'in the colonial army in the Sudan' was busy
'cluttering the house with antelope heads'. No doubt the property also
served as an unofficial museum, showcasing Dawson's eclectic tastes and, by
now extensive, antiquarian collection of artefacts and oddities.

Having been removed from Castle Lodge, the Sussex Archaeological
Society finally managed to locate a secure base for their offices, library and
museum within Barbican House, a three-storey town house on the eastern
edge of Castle Precincts. The property sat at the southern end of Castle
Precincts, directly opposite the Dawson family home. Thus the relocation
from Uckfield now meant that the Dawsons, as they sallied forth in to the
Lewes High Street from their family home, were to encounter 'the daily
coolness of the recently evicted tenants'. This cannot have been easy, for the
society members numbered hundreds and it is clear that they all blamed
Charles Dawson for the recent spate of unfortunate events, although the
thick-skinned Dawson was evidently able to cope.

Eventually, the move to Barbican House was to prove beneficial for the
Society, providing them with a slightly better-appointed property with
direct access to the High Street, but their relationship with Dawson was
never to recover. Even in the late 1940s, some thirty years after Dawson's
death, emotions towards the solicitor still ran high, the president Louis
Salzman commenting acerbically that 'his name was later given to the Pilt
Down Man (*Eoanthropus dawsoni*), the lowest known form of human being,
with the discovery of whose remains he was associated'.

Perhaps as a consequence of this new and rather awkward relationship with the county-based antiquarian group, Dawson re-established
links with his boyhood home, rejoining the Hastings Natural History
Society. Although both Charles and Helene were to remain fully paid-up
members of the Sussex Archaeological Society until Dawson's premature
death in 1916, neither appears to have played a significant or active role
within the organisation. Joining the Hastings Natural History Society
may well have been a perfectly natural step for Dawson to have taken in

any case, for he had already been a co-founder of the Hastings Museum Association and had an active interest, not to say passion, in the history of the town's castle.

The Bricks

In 1906, Dawson planned his most audacious fraud yet. This new endeav-our would involve the manufacture of an archaeological facsimile and its placement within a major, ongoing excavation. This particular ruse, if all went to plan, would not only increase Dawson's standing in the academic world, establishing him as a celebrity antiquarian *par excellence*, but would also simultaneously belittle all opponents within the Sussex Archaeological Society, effectively silencing them forever. It was a daring and inventive fraud, and one that bore all the hallmarks of his earlier escapades including an unwitting scientific dupe, vague provenance and a significant time lapse between initial discovery and final reporting of artefacts.

On Thursday 11 April 1907, Charles Dawson addressed a meeting of the Society of Antiquaries in London with a (literally) brand-new discovery: 'I have the honour,' Dawson stated proudly, 'to exhibit to the Society cer-tain impressed or stamped bricks and tiles, discovered by me in the Roman *Castra* at Pevensey, which have a bearing upon the date of the building of its walls.' The two pieces of brick and tile that he displayed to the society were stamped with the simple text 'HON AVG ANDRIA'. This, Dawson went on to suggest, could be translated as '*Honorius Augustus Anderida*'. These bricks not only had a bearing on the date of the Roman fortress at Pevensey, but also for the whole chronology of Roman Britain for they appeared to indicate nothing less than the last official building project authorised by the Roman state within the province of *Britannia*.

Flavius Honorius Augustus was emperor of the western half of the Roman Empire between AD 395 and 423 (his brother Arcadius ruling jointly in the east from Constantinople). It is widely acknowledged that his reign was not altogether successful; his uninspired leadership during the invasion of Italy by the Visigoths and subsequent sacking of Rome in AD 410 seriously destabilising the western half of the Mediterranean. Crucially, from the point of view of Roman Britain, it was Honorius who finally severed all official links with the province, following a series of rebel-lions there between AD 406 and 409. The discovery of tiles with his name on from a Roman fortress in Britain would, therefore, suggest a final piece of state-sanctioned garrison strengthening, prior to the first revolt against his rule in AD 406. No other inscription recovered from Britain had been made so late in the island's Roman history, and certainly no other text made such a strong link with the last legitimate Roman emperor to rule over the province. These pieces marked nothing less than the transition from Roman Britain to Saxon England.

What made things even more exciting, certainly from the perspective of the wider academic community, was that Dawson's bricks could actually be tied to a specific historical event. In AD 396 the central government of the emperor finally turned its attention towards the deteriorating military situation in Britain and sent a last expedition to resolve matters. The general in charge of the campaign was Honorius' chief of staff, Flavius Stilicho. Frustratingly, we do not possess a full and detailed account of the expedition, the only literary sources for it being the rather sycophantic verses of the Roman court poet Claudian who, in AD 398, noted 'with the Saxons subjugated the sea is now more peaceful, with the Picts broken Britain is secure'. Later, in AD 400, Claudian recorded that Britannia 'when on the point of death at the hands of neighbouring tribes, found in Stilicho protection'. Both references, though vague, would suggest a successful campaign around Britain's coastline, at some point between AD 398 and 400. Dawson's brick would easily fit within such a period of intense military activity.

Although the 'HON AVG' part of the inscription was easy enough to translate, the term 'ANDRIA' was more tricky. Dawson, however, confidently asserted that 'it suggests *Anderida, Anderesium* or *Andredes-ceaster*, names already identified with the *Castra* of Pevensey'. The tile therefore provided, in three short abbreviated words, an approximate date, an imperial sponsor and a name for the fort. There was no doubt about it: the bricks that Dawson had found at Pevensey were a major find, probably the most important discovery from Roman Britain.

The brick bearing the most complete inscription, Dawson noted in his talk and subsequent paper, had been found beneath the arch of the northern postern gate of the Roman wall in 1902. No one at the time seemed all that concerned that it had taken Dawson some *five years* to report the find, despite its clear importance. Crucially, once again, it was this time lapse that was to prove so important to Dawson in the establishment of the hoax, allowing him to be subtly vague about both the circumstances and context of his find, whilst also achieving primacy over the Sussex Archaeological

Society who had, since 1906, been clearing an area of the interior to the immediate south of the Roman postern.

The brick had, Dawson believed 'fallen down with other pieces from the roof of the arch, where similarly burnt bricks are to be seen'. Close examination of the remaining in situ pieces did not, however, reveal any further examples either of the same fabric or bearing any similar form of inscribed stamp. In a footnote to the published article, however, Dawson noted: 'I have also found portions of red brick from the eastern part of the wall bearing the mutilated outline of the same stamp.' These additional pieces in the eastern wall gave further credence to Dawson's finds, suggesting that the late Roman repairs undertaken by the government of Honorius had originally been more extensive. Unfortunately, these supplementary pieces of brick with the mutilated stamp were never mentioned again by the solicitor and no one, either at or after the London meeting, thought to press him on their precise number and location within the eastern wall of the fortress. This was unfortunate for, if they had, the hoax may well have started to unravel, for it would appear that the pieces existed solely in Dawson's imagination.

The key to the hoax, what gave it any sort of credibility, was that Dawson had, by the time of his talk to the Society of Antiquaries, successfully planted a third brick in the 1906 excavations conducted by the Sussex Archaeological Society at Pevensey. The impressed ceramic in question was in a broken state, the Latin text being incomplete and difficult to decipher. After much work, the chief director of the fieldwork, Louis Salzman, a man it should be remembered, who had already fallen out with Dawson over the sale of Castle Lodge in Lewes and the subsequent eviction of the *Sussex Archaeological Society*, suggested that the brick stamp had read ' ... ON AVG ... NDR ... '. The 'AVG' element suggested part of an imperial title, AUGUSTUS, but the rest of the text was difficult to decode, the fragmentary state of the brick not lending itself to any wild leap of interpretative logic.

In the final excavation report, published a year after Dawson had made his excited presentation to the Society of Antiquaries, Salzman singularly failed to provide any indication as to where, within his excavation, the broken brick had been located. Perhaps Dawson had planted the find in the general area of the Sussex Archaeological Society's dig, so that it could easily be found by a member of the team. The dig itself had not been screened from the public, work being conducted in an area that was freely accessible,

night or day, so the planting of artefacts, either in the general area of operations, topsoil or spoil heap, or within the finds assemblage, would not have
been difficult. Whatever the nature of delivery, the brick was successfully
secreted within Salzman's area of operations so that it could be 'found' by a
member of his team. Now that the stamped brick had entered the archaeological record as a genuine discovery made by in the context of a legitimate
excavation, Dawson could bring forward the next stage of his hoax.

The Honorius brick was a find that was clearly exceptional. For such
a well-preserved piece to be discovered, purely by chance, on the ground
by anyone involved in or visiting the Sussex Archaeological Society's dig,
would not have been impossible. For the 'find' to have been made by
Charles Dawson, who by 1906 was already known as the 'Wizard of Sussex'
on account of his amazing archaeological and geological discoveries, was
something that could elicit surprise, if not downright scepticism, within the
local antiquarian community. This was something that Dawson needed to
avoid at all costs, if his hoax were to succeed. The independent verification
of his 'discovery' by another was a masterstroke; Salzman's find helping to
authenticate Dawson's, the full version of the text on the later finds confirming (and simultaneously helping to interpret) the incomplete nature of
the first. Any suspicion surrounding the reliability or veracity of Dawson's
'finds' would, therefore, be likely to dissipate.

Salzman did not make very much of the brick; how could he? The fragmentary nature of the first piece ensured that, without a more complete
version of the text, his team could not decipher the inscription nor venture
a plausible interpretation. In fact, his thunder was totally stolen by Dawson,
who presented his bricks in a public lecture *and* subsequently managed
to publish his own article in the pages of the internationally renowned
Proceedings of the Society of Antiquaries, at least a year *before* Salzman's paper on
the dig (and his 'find') could appear in the pages of the *Sussex Archaeological
Collections*. Perhaps this explains Salzman's cursory treatment of the artefact
in his report where he notes, in hindsight almost grudgingly, that the translation of his find was only made possible through a 'comparison with a perfect
example from the same stamp, in the possession of Mr. Charles Dawson'.

The significance of the Honorius bricks was not lost upon the archaeological community and the pieces were referred to in all major discussions
of Roman Britain (all reports except, that is, those of the 1906–07 site
director of Pevensey, Louis Salzman, who only ever referred to them
once). Dawson had his spectacular find, his important 'transitional' artefact,

conferring further celebrity upon him; another spectacular discovery from the veritable Wizard of Sussex.

Doubts concerning the provenance and authenticity of the Pevensey bricks, however, began to surface throughout the late 1960s. The biggest worry for Roman specialists was the 'spidery' nature of the impressed lettering, a style that was quite unlike other official Roman military stamps, whilst the quartz sand fabric of the bricks did not match with any of those still preserved within the walls of the Roman fort. These doubts were expressed by John Mainwaring Baines, curator of Hastings Museum, to David Peacock of Southampton University early in the 1970s. Intrigued, Peacock decided to subject the bricks to a new form of scientific dating, namely thermoluminescence.

Thermoluminescence (or TL) is a method of dating geological samples, such as lava flows and meteorite impact craters, and archaeological materials, such as bricks, ceramics, hearths, kilns and all forms of heat-processed material (especially the residue of pyrotechnological or industrial activity). At its heart is the principle that minerals, when artificially heated, emit a flash of light, the intensity of which is proportional to the amount of radiation the sample has been exposed to as well as the length of time since the sample itself was last significantly heated. For thermoluminescence dating, the so-called mineral 'clock-resetting event' that will provide an estimate of when the ceramic brick or tile in question was originally fired, is heating to a temperature above 400 degrees centigrade. Needless to say, TL is a dating technique that was certainly not available in the early years of the twentieth century when discoveries such as the Pevensey bricks were first reported.

When the results of the TL dating of the Pevensey bricks were returned, the dates were startling. Dr S.J. Fleming of the Research Laboratory for Archaeology and the History of Art at the University of Oxford confidently calculated that the bricks possessed a 'firing date of no earlier than between 1900–1940 AD'. Retests and recalibrations were conducted: even the possibility that the wood glue that held the fragmentary Lewes example together may have affected the date was assessed. The results, however, remained stubbornly unchanging: the tiles were no older than the date at which they had been 'discovered'.

PHASE 2

A MAN OF ARTICLES

In the autumn of 1895, in recognition of his hard work and amazing discoveries at Lavant and Hastings, combined with his ever-increasing collection of archaeological curiosities, Charles Dawson had been elected a Fellow of the Society of Antiquaries, London. Now, as Charles Dawson FGS, FSA, his ambition changed again.

Dawson set his sights on a fellowship of the Royal Society, and he began to diversify, writing extensively on all aspects of archaeology (his spectacular 'finds' since 1895 had confirmed this) and history (both human and natural). He lectured, presented papers, organised exhibitions and published articles on subjects as diverse as palaeontology, ethnography, anthropology, chemistry, mineralogy, anatomy, heraldry, photography, lithics, ceramics, metallurgy, entomology, biology, aerodynamics and physics. He was becoming a true scholar; or rather, a jack of all trades and master of most. As a forger he had the knack of being able to identify just what material the various 'experts in the field' required in order to support their theories. He could identify the transitional phase or 'missing link' in most subject areas and 'discover' what academics had long thought really ought to be there. In short, he gave the academic community (and the public) what he thought they wanted.

His publication record to date, however, had been rather poor: two short articles in the *Sussex Archaeological Collections* for 1894 and a few scattered references within other people's work. This was something that the Uckfield solicitor instantly set out to rectify. Over the next few years Dawson's research and subsequent output was prodigious, especially when one considers that archaeology and history represented no more than a hobby, a distraction from his 'bread and butter' job in the legal profession.

The Papers

In 1898 Charles Dawson published a lengthy article entitled 'Ancient and Modern "Dene Holes" and their Makers' in the *Geological Magazine*. The term 'dene holes' was one which, in the late nineteenth century, covered a multitude of subterranean passages, tunnels and caverns found in the chalk of southern England. Few were well understood or dated and many were confused with Neolithic flint mines, Medieval quarries and a host of natural fissures or sinkholes. Dawson observed that, though a variety of theories concerning their origin had been made, there appeared little consensus as to function or purpose.

Dawson set out to resolve the many issues surrounding dene holes, 'by simple comparison … with excavations of exactly similar character and design'. He had, by his own account, explored 'two very fine' dene holes in Brighton together with John Lewis at some unspecified time in the past. We have only Dawson's word for this, however, it is fair to say that Lewis would have been the ideal partner on such an exploration, being a retired civil engineer who had previously been involved in underground construction work in India. Unfortunately, unlike his work at either the Lavant Caves or Hastings Castle, Lewis does not appear to have made a drawing of the subterranean workings beneath Brighton, or, if he did, these have not survived. He did, however, supply a section of a bell pit from Brightling in Sussex for the main plate accompanying Dawson's article, so there may have been some confusion at the editorial stage of the report as to which site was actually being described.

Dawson gives us very little detail concerning the shape or form of the Brighton/Brightling dene holes other than to note that one had been 'incomplete with respect to depth'. Removal of spoil from the subterranean tunnels appears to have been difficult, Dawson recording that larger blocks of rubble were hauled to the surface by a man operating 'a windlass of very primitive description' with a distinctively curved wooden handle. A sketch of this primitive windlass appears in Dawson's article and, although authorship is not credited, it possesses the distinctive style of John Lewis, presumably having been sketched at the time the two men made their exploration.

The rest of Dawson's article is concerned with his own personal observations of the Sussex dene holes, together with the results of interviews conducted with those involved in the digging and a lengthy description, compiled from other writers, of the dene holes of Hertfordshire and Kent 'made over a century ago'. In conclusion, Dawson noted that there was no real mystery attached to the pits, their purpose likely to have been solely for the extraction of chalk.

Shortly after its publication, a comment by an Essex-based geologist by the name of T.V. Holmes appeared in the *Geological Magazine*. Holmes was concerned that a report made by the Essex Field Club some ten years before had not been acknowledged in Dawson's article, something he felt was unfortunate given that some of Dawson's conclusions had already been reached in earlier publications. Holmes noted that a reader of Dawson's article could come away with the view that 'his bell-pit hypothesis is something quite new … whereas it was an old view before the report was written'.

In 1903, Dawson organised and co-ordinated a major exhibition in Lewes focusing upon the iron and pottery of Sussex. To accompany this he wrote a detailed article simply entitled 'Sussex Ironwork and Pottery', which was published in the *Sussex Archaeological Collections* for that year, together with a lengthy catalogue listing all objects displayed. A summary of Dawson's views on Sussex pottery was later published in *Antiquary*; entitled 'Sussex Pottery: a new classification'. It ended by urging that 'all who are interested should take a early opportunity of visiting the Lewes collection, since these specimens on loan must soon be dispersed and returned to their owners, perhaps never to be seen together again'. A second article by Dawson appeared in *Antiquary* two years later. This time, entitled 'Old Sussex Glass: its Origin and Decline', it purported to continue Dawson's research 'into the ancient industries of Sussex' whilst simultaneously hoping that additional material on the subject may be forthcoming.

Dawson's article on ironwork and pottery appears to be a weighty examination of both ancient Sussex industries, combining Dawson's own views with a sizeable section compiling the observations of earlier writers. One of the lengthier extracts quoted in support of Dawson's arguments (a piece from Dr Lardner's *Cabinet Cyclopaedia; Manufactures in Metal* of 1831) was supplied by the solicitor's colleague at Hastings Castle and the Lavant Caves, John Lewis. Lewis also provided a reconstruction drawing of a hammer forge taken from a painting of 1772 by the artist Joseph Wright. The publication was viewed by many as both the culmination of Dawson's extensive work into the subject area and a suitable tribute to the excellent exhibition organised and co-ordinated by him for the Sussex Archaeological Society. The ironworking section in particular was considered a major and notable contribution to the study of the defunct Wealden industry.

1. Charles Dawson in 1895. (Sussex Archaeological Society).

2. The Uckfield Urban District Council in 1897, shortly after Charles Dawson began his career in artefact fabrication. Dawson is standing in the centre row, third from right. His partner in Dawson Hart Solicitors, George Hart, is standing at the back, second from left.

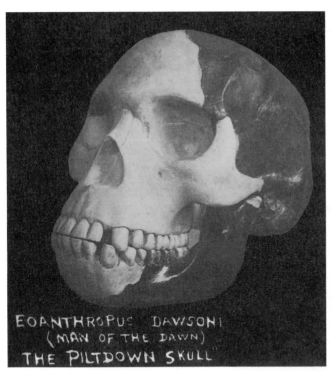

3. The skull of Piltdown Man (*Eoanthropus dawsoni*) as reconstructed from bone and teeth fragments 'found' between 1912 and 1914.

4. The missing link: A speculative recreation of Piltdown Man shaping an implement from a chunk of elephant bone with the aid of a crude flint tool or 'eolith'.

5. The unveiling of a commemorative monolith at the site of the Piltdown gravel pit by Sir Arthur Smith Woodward in 1938.

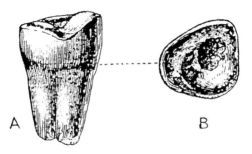

6. *Plagiaulax dawsoni*: two views of the first fossil tooth doctored by Charles Dawson in 1891.

7. The Lavant Caves: the partially backfilled brick entranceway built on the instructions of the Duke of Richmond.

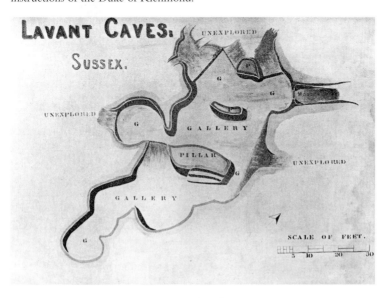

8. A plan of the Lavant Caves compiled by John Lewis.

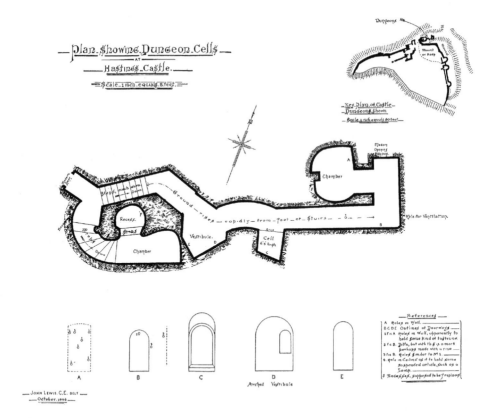

Plan. Showing. Dungeon. Cells
AT
Hastings Castle.

Scale. 1 inch. equals. 8 feet.

Key. Plan. of. Castle
Dungeons. Shown
Scale. 1 inch. equals. 80 feet.

Chamber

Hole for Ventilation.

Chamber

Vestibule

Cell
6 ft high

Recess

Steps

Ground rises rapidly from foot of Stairs

References
A Holes in Wall.
B.C.D.E Outlines of Doorways.
1 F is A Holes in Wall, apparently to hold some kind of fasteners.
2 F is B Ditto, but with this is a mark perhaps made with a ring.
3 F is B Holes similar to Nº 2.
4 Hole in Ceiling as if to hold some suspended article, such as a Lamp.
5 Recesses, supposed to be Fireplaces.

A   B   C   D   E
                  Arched Vestibule

John Lewis. C.E. delt
October. 1896

9. A plan of the Hastings Castle tunnels compiled by John Lewis.

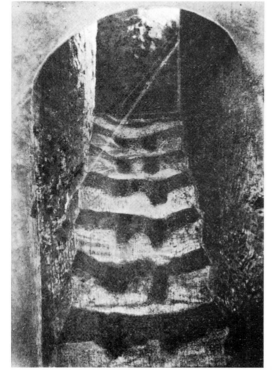

10. The heavily worn 'rock-hewn' steps of the Hastings Castle tunnel entrance, photographed by Charles Dawson in 1894.

DIAGRAM OF ORIGINAL PORTION OF SUPPOSED OCTAGONAL BASTION, WITH DOORWAYS TO DUNGEONS. (THE MODERN MASONRY IS OMITTED.)

PECULIAR MARKINGS, LIKE SHADOWS, FORMERLY ON THE SOUTH WALL OF THE MAIN GALLERY.

HASTINGS CASTLE DUNGEONS.

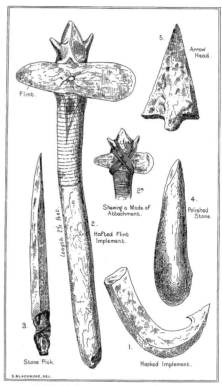

FLINT IMPLEMENTS FOUND NEAR EASTDEAN.

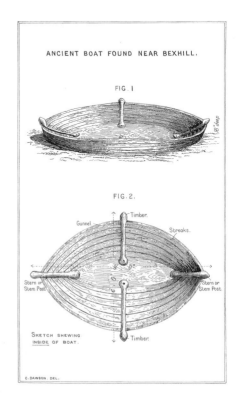

ANCIENT BOAT FOUND NEAR BEXHILL.

11 (above). A sketch compiled by Charles Dawson showing the entranceway to the Hastings Castle 'dungeons' and the shadow markings that he claimed to have seen on the southern wall of the northern tunnel.

12 (above right). A drawing of the main pieces in Stephen Blackmore's flint collection, as drawn for Dawson by John Lewis.

13. A drawing showing Dawson's reconstruction of the Bexhill boat.

14. The Beauport Park statuette, photographed by Charles Dawson in 1893.

15. The Uckfield horseshoe, drawn for the 1903 Lewes Exhibition.

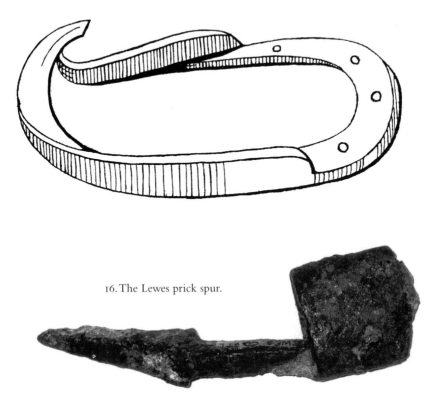

16. The Lewes prick spur.

17. The Bulverhythe antler hammer.

18. The 'socketed object' from the St Leonards Bronze Age hoard.

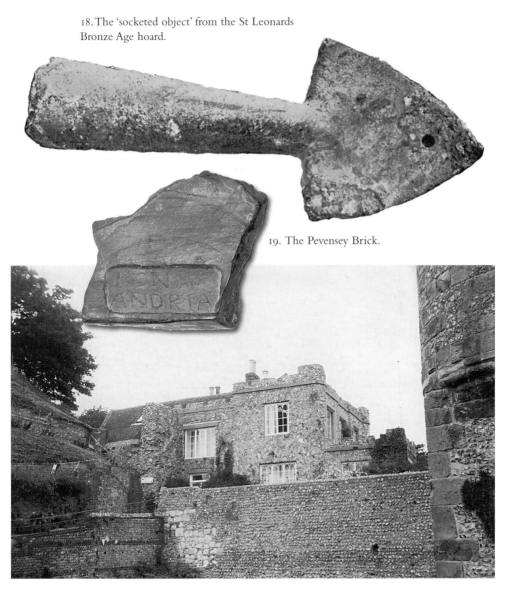

19. The Pevensey Brick.

20. Castle Lodge, Lewes, the Dawson family home nestled securely between the medieval barbican gate and castle motte.

21. Hastings Castle, a 'fish-eye' perspective photograph taken by Charles Dawson in 1909.

22. The Toad in a Hole.

In 1953, following the revelation of the Piltdown hoax, Robert Downes, who had made a study of the English iron industry, reread Dawson's work in the *Sussex Archaeological Collections*. Having minutely examined the work, Downes concluded that at least three quarters of the article has been plagiarised: 'having been taken directly from at least ten published sources, but mostly from William Topley's *Memoir on the Geology of the Weald*. This observation was picked up by Joseph Weiner, to whom Downes passed his conclusions, who used it in his book *The Piltdown Forgery*, noting that Topley's work had been shamelessly used 'almost word for word without acknowledgement'. Forty years later, John Walsh was to observe that 'all the usual tricks of the accomplished plagiarist' were present in Dawson's articles, including 'specific acknowledgements to a fact or two, when in reality whole sections had been lifted from the cited work'; the disassembly and 'paraphrasing of longer passages' taken wholesale from the earlier source; the 'subtle weaving together of facts and phrases from different texts'; and the 'incomplete or deliberately inaccurate acknowledgements'. Dawson's real contribution to the 'new work' was, in Walsh's view, nothing more than 'a few simple comments, some personal observations, and transitional material'.

The Bayeux Tapestry was to form the basis of Dawson's next foray into academic writing, discussing in the 1907 edition of the *Antiquary* how the piece had coped 'in the Hands of the Restorers'. The report, which critically assessed the nature and contribution of 'successive restorations and changes in conservation techniques', was prefaced by a note from Dawson, which stated that the his article 'represents a "by-product" of a critical study of the Tapestry and its literature, undertaken for a larger work dealing, in part, with the early phases of the Norman Conquest of England'. Dawson never published a 'larger work' on the early years of the Norman Conquest, so it is, perhaps, fair to assume that he was referring to his two volume analysis of Hastings Castle, which finally appeared in 1910.

No specific claims concerning the report's perceived authenticity (or lack of it) have ever been made, though Walsh, in his book *Piltdown Unravelled*, makes the comment ' for many years this article was cited in tapestry literature as an important contribution, and since no study of possible sources has yet been made, nothing can be ventured here about its origins'.

The Book

In 1910, Dawson's magnum opus, the *History of Hastings Castle: the Castlery, Rape and Battle of Hastings, to Which is Added a History of the Collegiate Church Within the Castle, and Its Prebends* was finally published in two volumes by Constable and Co. Ltd, of London. The work was supposed to have appeared a year earlier, indeed the title pages of both volumes still show the publication date as 1909, but some unspecified delay seems to have prevented this from happening.

The two volume *History* appears to represent a major academic endeavour, every piece of data concerning the fortress and its surroundings having been assembled, collated and compiled. In the preface to the work Dawson sets his reasons for having commenced such an obviously time-consuming study, observing that, although the castle was undoubtedly a famous and nationally significant monument 'no authentic record of its history exists'. Dawson, therefore, felt it imperative that the gap in knowledge be addressed and he decided to 'search out its records in the British and foreign depositories, public and private; and arrange them in chronological order, interspersed with extract from contemporary chronicles, in such a manner that the whole collection may tell its own story'.

The contemporary works which Dawson set out to use in order that the castle may tell its own story were, of course, set down in a variety of different forms, utilising different languages and modes of speech, across a broad span of time. Dawson conceded that, given the lack of special knowledge and opportunities provided for students of history, any attempt to reproduce the early documents in their original form would serve only to make reading impossible. Therefore, he would provide a modest translation of the source material.

A large number of sources were specifically cited by Dawson in his Preface to Volume I of the *History*. These included Francis Grose's *Antiquities of England and Wales*, 1773–87, Volume III (which Dawson noted 'may be considered the first important printed topographical reference'), William Herbert's 1824 letter-press to 'Moss's *History of Hastings*' (noted as 'the first serious attempt to unravel the history of the Castle') and G. T. Clarke's *Medieval Military Architecture of England*, 1884 ('an excellent if brief description of the architectural details of the Castle and Chapel'). Other materials used for the compilation of the *History* were, Dawson noted, the result of his own private research within the archives of the Public Record Office, the Royal Historical Commission, various foreign and diocesan records and other 'private sources'.

The *History of Hastings Castle* soon became a standard work and is still, on occasion, cited today. At the time of its publication, however, not everyone appeared so enthusiastic; one reviewer for the *Sussex Archaeological Collections* commenting that, though Dawson had 'displayed much industry in collecting material', he had exercised 'little judgement in its selection and arrangement', noting the many errors, inaccuracies and, ominously, the sections of text borrowed wholesale from other authors of which 'no acknowledgement of the source is made'.

Of course, by this time Charles Dawson had few, if any, friends in the Sussex Archaeological Society, his purchase of Castle Lodge (and subsequent eviction of the organisation) being an all too unpleasantly recent memory. In 1953, John Mainwaring Baines, then curator of Hastings Museum, compared Dawson's book with a copy of an unpublished manuscript on the castle written in 1824 by William Herbert. Baines later related he found both the manuscript and Dawson's *History* to be 'almost identical'. He reiterated his views to Joseph Weiner, adding more forcefully that Dawson's two volumes of the *History of Hastings Castle* had been 'copied unblushingly from Herbert's manuscript'; the remainder he accused of being 'gross padding'.

Following the revelation of the Piltdown hoax, Robert Downes reviewed the *History of Hastings Castle* to see if the 'inaccuracy and neglect' cited in the *Sussex Archaeological Collection* was actually evidence of more fraudulent activity. In a damning indictment of the two volumes, Downes confirmed Baines' view that the bulk of Dawson's work had been plagiarised from a variety of sources, the main one being Herbert's unpublished manuscript of 1824.

Dawson's next foray into publication was his analysis of the Red Mounds of Essex published in the 1911 volume of *Antiquary*. The red clay deposits in question occurred in large patches along the old tidal margins of the Essex salt marshes and were known locally as 'Saltings'. As with his earlier works, Dawson confided that much had already been written concerning these hills, the most informative research having been conducted by a group known as the Red Hills Exploration Committee.

The mounds or 'Red Hills' comprised burnt earth, wood, ash and slag or vitrified material. They appeared to represent the debris caused by a 'great extinct industry', Dawson mused, possibly appearing in Essex as the remains of ballast in ships 'sailing from some one or more great pottery centres'. Dawson assessed the evidence available for the mounds, concluding that they were probably of late Iron Age date. A response to the paper was sent from the aforementioned Red Hills Exploration Committee almost immediately, noting that the summary of evidence produced by Dawson was 'largely a repetition of what has already appeared in the reports', accusations that are curiously similar to those made by the Essex Field Club following the publication of their article on dene holes, which appeared in the *Geological Magazine*.

As we have seen, accusations of literary theft have frequently been made against Dawson for the articles that he wrote between 1898 and 1911. Certainly claims of naive referencing could be made for a great deal of his work, especially those publications where he loosely attributed the source or where he quoted whole chunks of text without accurately referencing it. In publishing, it is fair to say, that accusations of plagiarism can be made against a writer if it can be proved that they have taken or copied the writings, thoughts, ideas or inventions of another and passed them off as their own. Plagiarising may further be defined as the failure to properly reference or attribute pieces of work or quotations taken directly taken from an article, book or unpublished manuscript that is not the author's own.

When one looks carefully at Dawson's articles on Sussex ironwork and pottery, dene holes and the Red Hills of Essex, it can be seen that they contain a large amount of work attributable to other people. It must be noted, however, that these elements *are* referenced (albeit vaguely). All the lengthy quotations *are* sourced and all appear within inverted commas, something which helps distance them from Dawson's own text. In hindsight then, whilst it is difficult to specifically cite plagiarism in such instances, it is fair to say that the papers are, in reality, acts of compilation rather than original research. As such, Dawson should perhaps be more accurately credited as the 'editor' of these articles than the 'author'.

In this respect, the *History of Hastings Castle* represents Dawson's greatest act of compilation. Here he amassed an incredible wealth of primary data, secondary references and miscellanea, most of which was inaccessible to the ordinary reader, and republished them with his own comments, thoughts and conclusions. Downes is correct in stating that a large quantity of the *History* Parts II, III and IV was taken directly from Herbert's unpublished manuscript, but allowing the records to tell their own story is of course something which Dawson carefully and expressly cites as being the primary aim of the *History*. Perhaps, if Dawson had been credited as editor or compiler, rather than author, then the issue of plagiarism would never have arisen. As it is, the most serious accusation that may be brought against the solicitor in the case of his major publications is that he was not afraid of taking a short cut.

All of his works appear to have been hastily compiled, using extensive amounts of earlier work in the form of extended passages quoted almost verbatim. Dawson defended his action by stating that he wanted the original records to speak for themselves, but in reality he was desperate to get his name attached to as many publications, covering a variety of diverse subjects, as possible.

PHASE 3

A CURIOUS MIND

Charles Dawson's interests continued to diversify following his election to the prestigious Society of Antiquaries in 1895. Throughout the early years of the twentieth century, he became increasingly involved in the exploration of strange and unusual natural phenomena. He used his photographic talents to experiment with the recording of lightning. He discovered unusual forms of fish in local ponds (which he dutifully forwarded to his friend Arthur Smith Woodward), and examined genetic abnormalities in cart horses. He lectured at length about the discovery and properties of natural gas beneath the Sussex town of Heathfield. He observed toads preserved inside solid nodules of flint and discovered a 'new race' of humans from the icy lands of the Arctic Circle. He revived his earlier interests in fossil hunting, recovering additional (and freshly fabricated) evidence for *Plagiaulax dawsoni* and retrieving remains of *Iguanodon* and *Lepidotus* from Roar Quarry near Hastings for the Natural History Museum collection. He even managed to record a rare sighting of the English Channel Sea Serpent, a creature that had arrogantly defied the British public throughout the eighteenth and nineteenth centuries.

All these discoveries and experiments, a testament to what his friend and colleague Arthur Smith Woodward was later to call a 'restless mind, ever alert to note anything unusual', demonstrated Dawson's unparalleled knowledge and ever increasing expertise in a variety of diverse scientific fields. At the same time that these amazing observations were being made, however, the solicitor appears to have craved election to the Royal Society, something that would have marked the pinnacle of academic recognition. That his 'discoveries' not only continued during this period, but intensified, becoming more peculiar all the time, is probably not unconnected. Dawson's first candidacy certificate for Fellowship was filed on 19 December 1913, and was renewed every year, without success, until his death in 1916.

The Toad

One of the strangest of Charles Dawson's natural discoveries during this period was that of the so-called 'Toad in a Hole' presented to the Brighton and Hove Natural History and Philosophical Society on 18 April 1901. The artefact comprised an apparently mummified toad preserved within a hollow nodule of flint. The 'curiously light' lemon-shaped flint had, according to Dawson, been broken apart by two Lewes workmen, Joseph Isted and Thomas Nye, in the summer of 1898, and had been examined in situ by a Dr J. Burbridge. Quite how the artefact had then arrived in Dawson's hands was never fully established, though this could again be plausibly explained by the contact Dawson apparently kept with all workmen in his district. At the time of the presentation, the Lewes Toad in a Hole caused quite a stir amongst the scientific community and caught the popular imagination, with lengthy reports appearing in a number of papers including the *Illustrated London News*. Eventually, the artefact was passed on to Brighton's Booth Museum of Natural History, through the agency of Dawson's antiquarian colleague Henry Willett.

The Toad in a Hole is certainly a curious discovery, though it was one that was not entirely without precedent. In 1811, for example, the Derbyshire geologist White Watson reported that a doctor from Manchester University had, some years previously, broken apart a block of limestone only to discover a toad alive at the centre. In 1825, intrigued by the possibility that amphibians could survive for long periods within inhospitable environments, the naturalist Dr William Buckland conducted what is, by today's standards, a not altogether ethical experiment, by taking twenty-four toads placing them in 'sealed cells' and burying them in solid and porous limestone for just over a year. When the toads were exhumed, on 10 December 1826, Buckland noted, with some surprise, that only those fortunate enough to have been sealed within porous limestone had survived the experience. Though other cases of amphibian entombment in rock have occasionally been reported, Dawson's example was the first scientifically attested example. That alone should have rung alarm bells.

All the classic elements of a Dawson hoax were in play during the initial phases of the toad's 'discovery'. First, there was the uniqueness of the find itself: not in this instance providing an evolutionary or 'transitional' phase between species, but certainly establishing hard evidence for a popular rural myth. Secondly, there was the nature of the discovery: a random find made by ordinary workmen. This time Dawson supplied names, Joseph Isted and Thomas Nye, but no contact details. Even if Isted and Nye had existed, there was no way that any self-respecting scientific investigator could easily have made contact with them for the purposes of interview, for Dawson had employed his third tactic: the significant time delay. Once again, no one questioned why it had taken Dawson nearly *three years* to bring the find to public attention or, indeed, where precisely the toad had been in the period between discovery and reporting. Dawson could, of course, have passed the blame onto the mysterious Dr J. Burbridge who, he reports, first discussed the toad with the workmen. Perhaps Burbridge retained the find himself for a period, without realising its significance. Perhaps he is simply another example of a fictional character, created to lend credence to an unusual story. Certainly he is never mentioned again.

The fourth element in the hoax was the way in which the 'discovery' was reported, not via a refereed academic paper in a peer-reviewed scientific journal, but in the press, in this instance the *Illustrated London News*. Such first exposure was guaranteed to obtain maximum news coverage, boosting Dawson's own celebrity antiquarian status, without all the tedious need for writing a serious article. Serious articles required serious thinking, writing and referencing, with no guarantee of final acceptance in the chosen journal. They also required hard fact concerning the nature of the initial discovery, something that, if provided, could expose the weaknesses in Dawson's story. The fifth and final element in the successful hoax was the unwitting scientific dupe, in this case Dawson's antiquarian colleague Henry Willett, who appears to have had nothing to do with the toad, other than to act as intermediary between Dawson himself and the Booth Museum of Natural History, where the toad remains to this day.

Since its presentation to the museum in 1901, the toad has shrunk considerably, indicating not only that it was not 'mummified', as claimed, but also that it was unlikely to have been in the possession of either Dawson or Dr Burbridge for three years following exhumation, otherwise it would have certainly shrunk well before its formal accession to the museum.

The Serpent

Five years after his reporting of the Toad in a Hole, Charles Dawson encountered a second mystery of nature, this time a monster from the deep. Writing
to Arthur Smith Woodward on 7 October 1907, Dawson related that, whilst
travelling on board the steamer SS *Manche*, on what was otherwise a fairly
a routine journey from Newhaven to Dieppe, he had spotted something
strange through his binoculars. The creature glimpsed was, Dawson related,
'some two miles away' from the steamer and, whilst he had it in his sights,
the solicitor was joined by two other passengers. One, also glimpsing the
object breaking the surface, said 'Hallo! What's that coming, the sea-serpent
or what is it?'. Whatever it was, object shifted course, turning 45 degrees
from the port side, where it graciously provided 'a more extended and less
complicated view'.

The three passengers could not make out head or tail, but 'a series of very
rounded arched loops like the most conventional old sea-serpent you could
imagine' were clearly visible from their vantage point on deck. 'The loops,'
Dawson went on, 'were fully 8 feet out of the water, and the length 60 to
70 feet at the smallest computation.' Transfixed, he watched it receding from
the path of the ship until, as it came between the ship and the sun, it disappeared forever.

Fortunately, Dawson possessed his camera, a small Kodak, and was able
to take 'several shots' of the beast as it came about. Sadly, the developed film
showed no sea serpent or 'detail of the sea beyond a few yards'. Undeterred
by his unusually bad luck in photographic recording, Dawson explained to
Woodward that he had discussed the sighting in some detail with his fellow
passengers, exchanging cards with several of them in the hope of bring-
ing forth witness statements if required. There is no evidence that either
Dawson or Woodward ever followed up these witnesses to collect formal
statements. Furthermore, there is no evidence that Woodward ever questioned his friend over *why* it had taken him so long to report the sighting.
Dawson wrote to Woodward on 7 October 1907, but the sea serpent had
been seen on Good Friday 1906, a whole eighteen months *before* Dawson
finally put pen to paper.

Strange as it may seem today, the possible existence of sea serpents was, in
the later half of the nineteenth century, a popular topic of conversation. Sea
monsters are, of course, present in the mythology of most ancient cultures
and, despite the more extensive exploration of the world's oceans in the last

century, they remain a potent symbol of 'the great unknown'. The majority of monster sightings through the eighteenth and nineteenth century were, however, largely unsubstantiated or related by those considered to be wholly unreliable. All this changed for the great British public in 1848 when Peter M'Quhae, captain of Her Majesty's frigate HMS *Daedalus* reported a monster sighting to the Admiralty. M'Quhae claimed that, between the Cape of Good Hope and St Helena, *Daedalus* had for some twenty minutes been accompanied by a large sea creature, which 'held at the pace of twelve to fifteen miles per hour, apparently on some determined purpose'. The head of the beast, which M'Quhae claimed resembled that of a snake but with the mane of a horse, was raised some 4ft (1.2m) above the water; the creature as a whole measuring at least 60ft (18.3m) in length.

When Captain M'Quhae's report was published in the London *Times*, it caused a sensation. The somewhat unsettling feeling that there might be similar creatures skulking in British coastal waters sparked a flurry of worried letters to the national press and generated a wave of mild hysteria for those communities facing the English Channel. A sense of the palpable unease caused by the *Daedalus* incident is reflected in the experience of HMS *St Vincent*, which in the summer of 1848, only weeks after M'Quhae's report had been leaked to the press, was lying at Spithead anchorage, off Portsmouth in Hampshire. W. Gore Jones, then attached to the crew of the *St Vincent* later recalled that, one evening at about 6 p.m., just as the officers were starting dinner, word from the watch came that 'a sea serpent was passing rapidly between the ship and the Isle of Wight'. Up on deck, the officers identified what appeared to be 'a large monster, with a head and shaggy mane, about 100 feet long' swimming alongside with 'a rapid, undulating motion'. Boats were launched with armed marines in hot pursuit, but, on approaching the 'beast', it became apparent that the creature was in fact nothing more than 'a long line of soot' pitched from a passing steamer. Despite this wholly explicable 'creature' incident, sightings of sea serpents in the English Channel were to remain a regular fixture in the national press; letters to *The Times* continuing until the early 1890s.

The case of the English Channel sea serpent seen by Dawson is a particularly strange one, for the solicitor made no public acknowledgement of his sighting, other than in a private letter to his friend Woodward. Perhaps, given the nature of the evidence, a lack of confidence on Dawson's part may well not be surprising for, even today, those who claim to have witnessed unnatural events or mysterious creatures often find themselves outcasts from polite society. Dawson may have felt that his 'sea serpent' was, without the evidence of his camera, something that would not be taken seriously by either the media or the wider scientific community. Perhaps he was not sure himself about what he and his fellow passengers had actually seen that day. Perhaps they had seen a line of soot, similar to that encountered by the HMS *St Vincent* in 1848, or another form of Channel pollutant. Perhaps he made the whole story up and was just sounding Woodward out to ascertain whether he believed him; to see, in other words, whether Woodward, the curator at the Natural History Museum would make for a useful academic dupe in the future.

Unfortunately, Woodward's response to the whole affair is lost to us. Was he intrigued by the details supplied by Dawson or, as an expert on marine palaeontology, did he laugh the incident off as the misidentification of something else in the water? Even if Woodward had been interested enough to attempt to follow up potential witness statements, due to the lengthy gap between sighting and reporting (Woodward not getting to hear about the serpent until early in October 1907), it may well have proved difficult to locate Dawson's travelling companions, let alone extract a convincing report from them. This was another example of the inexplicable time delay at work.

What is, in retrospect, interesting is the form taken by Dawson's reported creature. His description of 'very rounded arched loops' does not sound like either the *Daedalus* or *St Vincent* sea serpents, but it does fit perfectly with the late nineteenth and early twentieth-century view of what it was felt a lake serpent, such as that often ascribed to Loch Ness in Scotland, *ought* to look like. Up until the late 1960s, the Loch Ness Monster (or 'Nessie') was often described or drawn (especially by countless school children) as a collection of coils and humps emerging from the surface of the Loch. From the 1970s, however, the general form of Nessie, and indeed of all sightings relating to her, changed. This is due in no small part to the work of Dr Robert Rines at Loch Ness between 1971 and 1975. Rines claimed, at this time, to have photographed a great beast with a long neck, wide body and diamond-shape flippers: in short, a Plesiosaur. Plesiosaurs, a type of marine reptile that inhabited freshwater and marine environments of the Triassic to Cretaceous periods, fed mainly on fish and other small water creatures such as ammonites. Since the publication of Rines' startling photographs of Nessie, people have only seen Plesiosaurs, the coiled serpent being consigned to oblivion.

Dawson's observation of the marine serpent was clearly of its time, and, given the general lack of sightings through the twentieth century, it does seem somewhat unlikely that anything quite as large as the creature that Dawson claimed to have seen could really be out in the English Channel. It has been shown that, when it comes to unexplained phenomena, people often see what they want to see. That is not to doubt that, in the case the world's oceans, there is not something large and wholly unknown lurking in the depths (in fact, given how little of the seabed has been adequately mapped, it would be surprising if there were not), but a large number of 'monster sightings' at somewhere like Loch Ness can be attributed to the human mind telling the eyes what really *ought* to be there: a Plesiosaur rather than, say, a piece of wood, an otter, a rock or a piece of loch debris.

This still begs the question: 'why did Dawson relate this curious story to Woodward in the first place?' Surely, if he were attempting to fool Woodward, there would appear to have been little to gain from such a deceit? Perhaps, however, it is the context of the alleged sighting that is important here, for in the early years of the twentieth century Dawson's attentions were shifting away from matters historical and archaeological, and more towards those of the natural world. Part of this shift may have been due to his election in 1895 to the Society of Antiquaries, in much the same way that his passion for geology seems to have declined following his election to the Geological Society in 1885. Having achieved such goals, Dawson may have felt that to continue in such fields would be akin to merely treading water whilst diversifying would prove an altogether more challenging experience.

Perhaps the increasingly poor reception that his work was receiving from Sussex antiquarians, or indeed the major falling out with the Sussex Archaeological Society following the Castle Lodge incident of 1904, helped convince him that his research should be directed elsewhere. We will never be entirely sure, but it is to this period that the more unorthodox aspects of the natural world, such as the Toad in a Hole, the 'incipient horns' of cart horses, the goldfish/carp cross and wholly new species of human really began to dominate his time.

The New Man

On 12 May, 1912, barely one month after Dawson alerted Arthur Smith Woodward to the first discovery of what was to become Piltdown Man, the solicitor from Sussex was again writing to Woodward, this time about the discovery of an entirely new race of human. Not content with the developing Piltdown hoax, still in its early stages, it would seem that Dawson was still casting around for the 'big discovery' in the hope of enhancing his prospect of gaining a fellowship of the Royal Society.

'Since I saw you,' Dawson commented in his letter, 'I have been writing on the subject of "The 13th Dorsal Vertebra" in certain human skeletons, which I believe is a new subject.' It was not a new subject, as the solicitor himself well knew, but it was an area in which Dawson clearly wanted primacy. 'I send you the result,' he went on, 'and if you think well enough of it I should be very much obliged if you would introduce the paper for me at the Royal Society.' Dawson obviously felt that his discovery was of sufficient importance to be reported directly to the Royal Society, though, as he had not been elected to a fellowship, the solicitor could not himself speak to the group. In 1912, however, Arthur Smith Woodward was already a fellow (rising to become a member of its council the following year), and was, therefore, in an excellent position to introduce his friend's paper to the group.

The original draft of Dawson's report, alluded to in this letter, entitled 'On the persistence of a 13th Dorsal Vertebra in Certain Human Races', is preserved in the Library of Palaeontology at the Natural History Museum in London. It deals with the observation that, though the normal number of vertebrae in a human skeleton is twelve, certain Inuit skeletons in the Royal College of Surgeons appear to possess thirteen. An extra vertebra is common enough in apes and the loss of this during the process of human evolution was often ascribed to the development of an upright walking posture. Human skeletons retaining that '13th Dorsal Vertebra' were, therefore, an anomaly, and one that Dawson appeared keen to explain.

Curiously, Dawson added an element of conspiracy to his letter, suggesting that others were aware of his discovery and were on the verge of supplanting him. 'I am very anxious,' he went on, 'to get it placed at once because I have had to work the photographs under the nose of Keith and his assistant.' Dawson's original paper is accompanied by a number of photographs of 'skeletons displayed in glass cabinets in the museum exhibition halls of the Royal College of Surgeons' where Arthur Keith worked as conservator. The photographs appear to have been taken hastily, some being 'marred by the glare of the flash', which may support Dawson's comment that they were taken 'under the nose of Keith and his assistant'.

The assistant went unnamed, but Dawson alluded to the urgency of the work, adding, 'I want to secure the priority to which I am entitled'. In other words, Dawson wanted to present his unique theories to the Royal Society as quickly as possible.

The urgency with which Dawson asks Woodward to press ahead with his draft paper appears, in the letter of 12 May, to have stemmed from his belief that his theories and observations will be taken from him by Keith or his assistant once the true significance of his findings became known (although quite why, as anatomists, neither had noticed the anomaly before, especially in the skeletons on display in their own college, is never satisfactorily explained). Of course, the real reason for Dawson's appeal for urgency was that he required swift access to the higher echelons of the Royal Society before anyone realised that his research was not, in fact, new.

In February 1912, a book had been published by the French anatomist A.F. Le Double entitled *Variations de la colonne vertbrale de l'homme, et leur signification au point de vue de l'anthropolie zoologique*. The tome, which covered the same basic themes as those propounded by Dawson, had received little attention in Britain, although we know that Arthur Keith had been forwarded a copy by the Royal Anthropological Institute who were keen for him to write a review. By May, Le Double's book, although circulated to some limited degree in Britain, had yet to receive a full English translation. Dawson was fluent in French, and may, therefore, have been aware of the existence of both book and content before many of his scientific associates. As such his sense of urgency to 'secure the priority' to which he felt entitled may reflect the very real worry that Woodward, Keith, or any other member of the Royal Society might hear of Le Double's research before they were aware of his. Certainly, the announcement of Dawson's 'discovery', coming so soon *after* the publication of Le Double's book (which itself took years to fully compile), casts severe doubt over the reliability of the solicitor's research.

Perhaps because of the growing awareness of the French anthropologist's work and the wider dissemination of his book, Dawson's paper never saw the light of day, being consigned to the archives of the Natural History Museum. Perhaps Woodward himself was simply not interested (his reply, if there was one, to Dawson's letter has not survived) or perhaps the excitement of the Piltdown discoveries took off in a very unexpected way, making Dawson decide to focus his time on the elaboration of this particular fraud, rather than pursuing the thirteenth dorsal vertebra. The British press were eventually to run elements of the story 'as the discovery of a new race by the discoverer of Piltdown', but the academic impact of Dawson's unpublished article was ultimately negligible. By the summer of 1912, Dawson had far more important things to be concerned about.

PHASE 4

THE BIG DISCOVERY

With the fellowship of the Royal Society now firmly in his sights, Charles Dawson continued his investigations of the more unusual and unorthodox aspects of the natural world. Dawson also reinvigorated his former interest in palaeontology, finding more (faked) teeth of *Plagiaulax dawsoni* and, working with a number of amateur geologists from France, discovering the remains of at least two more unique mammals.

None of these 'discoveries' seemed to be taking him anywhere, however, and by March 1909 Dawson wrote to his friend Arthur Smith Woodward complaining that he was 'waiting for the big "find" which never seems to come along'.

The Ape Man

Late in 1911, a chance meeting with the writer Sir Arthur Conan Doyle was to prove the inspiration for Dawson's most daring and ingenious hoax: at last, the Sussex solicitor was on the verge of making his 'big find' and to do so he would marshal all of his skills in fraud, fabrication and deceit. This deception, if it all went to plan, would not only gain him greater notoriety and celebrity, but also win him both a fellowship of the Royal Society and, just perhaps, a knighthood too.

Many investigators into the Piltdown story have noted the apparent coincidence of timing between Arthur Conan Doyle's 1912 novel *The Lost World*, complete with its tribes of hairy ape-men, and the first announcement to the world of Piltdown Man, the missing link between apes and humans. Given that, at the time of the discovery, Doyle was living relatively close to the Barkham Manor site, where Dawson claims to have found evidence for his early man, has convinced some that there must be a link between the novelist and the hoax; that even if Doyle had not conceived the fraud, that he must somehow have been aware of it, perhaps even trying (belatedly) to expose it in the pages of his novel.

Arthur Conan Doyle was born in 1859 in Edinburgh. He had trained as a doctor, graduating from Edinburgh University in 1881 but, after setting up practice in Plymouth then Portsmouth, Doyle began writing fiction. His most famous creation, the detective Sherlock Holmes, first appeared in the *Beeton's Christmas Annual* for December 1886. A series of increasingly successful short stories and other novels featuring Holmes followed and by 1891 Doyle had abandoned his career in the medical profession to become a full-time writer. In 1907 he married Jean Elizabeth Leckie and the two moved to Windlesham House at the margins of Crowborough, a small town in East Sussex.

Certainly the publication of Doyle's book, *The Lost World*, so close to the announcement of Dawson's discovery appears fortuitous and may hint that the author knew more about the forgery than he let on. When one considers the exact chronology and sequence of events surrounding both the emergence of the book and the announcement of Piltdown Man, however, a new and more interesting scenario presents itself.

That Arthur Conan Doyle knew Charles Dawson is apparent enough, but the exact circumstances behind their friendship remain unclear. Their first confirmed meeting occurred late in 1911, something established by correspondence surviving in the archives of the Natural History Museum, although they may well have met as early as 1909, following an inquiry made by Doyle to Arthur Woodward concerning dinosaur footprints discovered near Crowborough. Woodward had written to Dawson, asking if it were possible to visit the celebrity author, who had for sometime been corresponding with the museum, in the hope of determining what exactly it was that he had found.

By November 1911, the Doyles and the Dawsons were on sufficiently good terms for Charles and his wife Helene to be invited to Windlesham for lunch. After the meal, Dawson examined a number of Doyle's discoveries. Unfortunately, the 'great fossil' that the author had been most enthusiastic about proved to be nothing more than a 'concretion of oxide of iron and sand', though the discovery of a Neolithic flint arrowhead by Dawson 'in view of us all' apparently saved the day. Given the context and timing of the lunch, it seems inescapable that Doyle must at some point have mentioned his latest novel *The Lost World*, the writing of which he had by then completed. What Dawson made of the book we do not know though he later recorded in a letter to Arthur Woodward that it was set 'on some wonderful plateau in S. America with a lake which somehow got isolated from "Oolitic" times'. The valley possessed all 'the fauna and flora of that period' Dawson went on 'and was visited by the usual "Professor"'. Characteristically, Dawson ended with the comment 'I hope someone has sorted out his fossils for him!'

The reality, of course, is that the relaxed discussion at Windlesham was Dawson's moment of revelation; the realisation that he had his 'big find'. Much of his antiquarian career to date had been spent providing people in the academic community with 'transitional' artefacts; missing links in the chain of technological development. Only once before, in 1891, with *Plagiaulax dawsoni*, had the solicitor generated an evolutionary 'missing link', in this case between reptiles and mammals. The hoax, involving the simple doctoring of fossil teeth, had been relatively low key but it had successfully deceived the geological community on at least two separate occasions, the second deception being conducted in 1906. As he sat listening to Doyle over lunch, the solicitor must have wondered whether he could now create the greatest deceit of all. Using technical skills honed over decades (such as the filing of *Plagiaulax* teeth and the whittling of the Bulverhythe antler) and his, by now watertight, *modus operandi* (involving vague provenance, initial discovery by unnamed workmen and time-lag in reporting), could Charles Dawson really give British science what it had craved for so long: a primitive human ancestor; a missing link from the Home Counties?

Between November 1911 and February of the following year, Dawson worked feverishly on his creation. To make this particular hoax work, he would have to ensure that everything was planned down to the last detail. He would require key archaeological components, the basic artefacts that represented the building blocks of the fraud, but he would also need to ensure there was a plausible 'discovery' story in place (one which he would have to stick to throughout any subsequent investigation), and that sufficient numbers of 'tame scientists', including prominent members of the academic community, were convinced and supportive right from the start.

Dawson's plan was relatively simple: he would, over a period of time, produce evidence that would conclusively prove the existence of a British 'missing link'. To do this successfully, Dawson had to move slowly. He could not risk exposure by producing all the finds in a relatively short amount of time, and, given his earlier success rate in unusual artefact retrieval, he would have to ensure that other specialists made key discoveries themselves. He would have to play the long game, with pieces of the puzzle coming to light over many months, if not years, and he would have to modify and adapt the strategy as new theories and questions arose. It would be dangerous, given that exposure would finish his career not only as an amateur antiquarian, casting all his earlier discoveries in doubt, but it would also end his life as a solicitor and all-round pillar of the community.

Locating a designated 'find spot' for fraud was relatively easy, for Dawson knew the landscape of East Sussex well and, as a solicitor, he acted as steward to a number of large and prosperous estates including the Manors of Barkham, Netherall and Camois. What he needed was an apparently secure sequence of deposits of the right age for early humans, a reason for their disturbance (ideally road digging or quarrying) that would explain how the finds were noticed in the first place, and a landowner who was more than happy to allow additional excavation and investigation. Barkham Manor, to the south of the Sussex village of Piltdown, was ideal in this respect, for Dawson knew both the area and the owners well, having very good reasons to visit in both a professional and social capacity. The manor sat over an extensive sequence of waterborne gravel, which was regularly quarried for localised road and pot-hole repair.

Dawson also needed to ensure that his find fitted the expectations of the scientific community. To some extent, he gambled, a *British* discovery would blind most experts to any inconsistencies in the execution of the fraud, so desperate were many to find evidence of human evolution in their own backyard, but the find would still have to be plausible. The features that were to define 'Piltdown Man', as Dawson's fossil finds were to become known, typified the early Edwardian view of human ancestry: an enlarged and well-developed cranium, emphasising powerful thought processes, combined with a more primitive, ape-like face. Increasing numbers of fossil discoveries made throughout the middle of the twentieth century, long after Dawson had died, demonstrated that the features that most clearly defined Piltdown, namely its ape-like jaw and teeth attached to a human forehead, were patently not present elsewhere; all the evidence indicating that it was the human jaw, supporting human-like teeth, that was a remarkably early feature in the development of *Homo*, whereas forehead and brain changed more gradually. Piltdown had these key features in reverse, but Dawson was not to know this at the time.

As concerns the means to acquire the raw materials necessary to furnish the fraud, a man such as Dawson would in fact have no difficulty at all in procuring 'unusual' items such as a human cranium and an ape-mandible'. In fact, Dawson would have found many of the basic building blocks necessary to accomplish the forgery readily to hand in Sussex. Dawson was a co-founder, in 1889, of the Hastings and St Leonards Museum Association and also a member of the museum's committee, in charge of the acquisition of antiquarian artefacts (such as Stephen Blackmore's flints) and historical documents. Indeed the museum itself regularly proved to be the ideal display case for Dawson's own extensive collection of antiquarian curiosities, which eventually occupied an entire section all of its own.

Whilst performing his legal role, as partner in Dawson and Hart, Dawson also found himself acting as solicitor to a number of prominent Sussex antiquarian collectors, and was able to catalogue a variety of materials bequeathed or otherwise donated to Hastings Museum throughout the 1890s and early 1900s. By the 1890s, Dawson was even conducting his own excavations in and around Hastings Castle, an early phase of which was reported to have produced a 'great haul' of artefacts. In addition to all this, we know that the solicitor was purchasing materials from local dealers (such as a ceremonial mace head bought from a pawnbroker's in Kent). We should also not forget that Dawson claimed to be in contact with all the workmen in his area 'who might make accidental discoveries'; contacts allegedly helping him to collect the items such as the Toad in Hole from Brighton. A man such as Dawson would certainly, therefore, be in a excellent position to accumulate the necessary specimens for any number of frauds, forgeries and deceptions.

Some have doubted that a 'mere country solicitor' would have had either the time to perpetrate the fraud, or the opportunity to acquire certain key ingredients. Time does not appear to have been a problem for Dawson, however, for the quantity of publications, exhibitions, displays, excavations, surveys and lectures that we know that he organised undermines the suggestion that he had little time off from work to feed his antiquarian habit. Quite how his schedule at Dawson and Hart operated is unclear, though it may have been that Dawson's partner there, George Hart, later to become

'Official Solicitor of England', shouldered most of the burden. Perhaps Dawson was simply a workaholic who never had time for normal 'everyday' activities.

By February 1912, Dawson was ready, initiating stage one of what was to be his most famous creation. Writing to Arthur Smith Woodward, belatedly appraising the London curator on the contents of Arthur Conan Doyle's new novel, which was to begin serialisation in *The Strand Magazine* in April of that year, Dawson dropped in a comment that must have instantly grabbed Woodward's attention. 'I have come across a very old Pleistocene (?) bed overlying the Hastings bed between Uckfield and Crowborough,' he stated casually, 'which I think is going to be interesting.' 'It has a lot of iron-stained flints in it,' he went on, 'so I suppose it is the oldest known flint gravel in the Weald.' Relating to other finds recovered from the layer, Dawson observed that he had retrieved '1 portion of a human skull which will rival *H. Heidelbergensis* in solidity'. The phrasing was garbled, possibly due to the solicitor's evident enthusiasm, but he appeared to be saying that he had actually *found* part of a human skull within this ancient deposit of gravel.

Dawson's nonchalant comparison with '*H. Heidelbergensis*' was a clever way of drawing the London curator into his trap. By 1912 *Homo heidelbergensis*, or Heidelberg Man, was considered to have been one of the earliest European human ancestors. Found in a quarry on the banks of the River Neckar near the village of Mauer (halfway between Heidelberg and Mannheim in Germany), the remains comprised a single robust, ape-like jaw, lacking a prominent chin. The jaw retained all of its teeth, which were quite human in appearance, along the right side, together with the first and second incisor, canine and third molar on the left. The bone, which derived from a deeply buried layer of sand and gravel, had been associated with a variety of fossilised mammal remains. Despite the relatively poor survival, Otto Schettensack, a palaeontologist from the University of Heidelberg, confidently asserted that the remains represented the best evidence to date for an ape-like ancestor of the human race; a predecessor of both modern *Homo sapiens* and the Neanderthals.

Dawson had, at some time prior to 1912, examined a cast of the famous Heidelberg jaw for himself. The exact date and circumstances of this examination are not recorded, though Dawson later related in the pages of the *Hastings and East Sussex Naturalist* that 'the massive appearance of the two pieces of cranium from Piltdown made it seem likely that they had belonged to an individual allied to the original possessor of the Heidelberg jaw'.

The retrieval of human remains from a 'very old Pleistocene bed' in the Weald was intriguing. Dawson had deliberately said very little in his letter, allowing Woodward to draw his own conclusions. If the skull fragments that he had recovered were in any way comparable to the Heidelberg find, this would undoubtedly prove to be an explosive discovery. Up to 1912, it is fair to say that British palaeontologists had been lagging behind their European counterparts. French and German scientists had, throughout the nineteenth century, produced a range of well-preserved Palaeolithic deposits including skeletal remains such as *Homo neanderthalensis* (Neanderthal man) and a wealth of worked flint assemblages, each helping to identify specific 'cultures' such as the Acheulean, Magdalenian, Levaloisian and Mousterian. In comparison, very little had been recovered from Britain, despite the best efforts of many capable palaeontologists, geologists and antiquarian researchers. To make matters worse, what meagre remains of the period had been recorded from British contexts provided some French palaeontologists with a rather derogatory term for their British colleagues: '*chasseurs de cailloux*' or 'pebble hunters'.

However eager Woodward may have been to see Dawson's find for himself, the workload of both men appears to have prevented either a meeting or an inspection of the material. On 24 March 1912, Dawson wrote to Woodward explaining that the proposed visit to the gravel bed at Piltdown 'will depend on the weather, at present the roads leading to it are impassable and excavation is out of the question'. Slowly but surely Dawson was reeling Woodward in. Two days later, on 26 March, Dawson forwarded two new finds from the gravel together with a brief note which read 'will you very kindly identify enclosed for me. I think the larger one is hippo?'. Once again, Dawson was being disingenuous, for he knew exactly what the 'larger one' was, having procured it himself and he gambled that it was just the thing that would excite Woodward's interest. Almost by return of post, the London curator replied with the note '28th Premolar of Hippopotamus'.

His appetite clearly whetted by Dawson's description of the human remains and their association with ancient mammals, Woodward counselled his friend to secrecy. On 28 March Dawson wrote: 'I will of course take care that no one sees the pieces of skull who has any knowledge of the subject and leave all to you.' He knew now that he had his man. 'I have decided to wait,' he wrote, 'until you and I can go over by ourselves to look at the bed of gravel. It is not far to walk from Uckfield and it will do us good!'

But Woodward could still not get sufficient time off from his job in the museum and Dawson's enthusiasm (and no doubt frustration), got the better of him. Despite his insistence to Woodward that the find would remain secret, he began showing his skull fragment to people. Perhaps the length of time it was taking for Woodward to visit was proving unbearable, or perhaps, more likely, Dawson was trying to stir up some attention to his find, in the hope of gathering a team of experts for any future expedition to the gravel bed. At this stage of the hoax, there was no guarantee that the London curator was going to be the pliable academic dupe that Dawson needed if his plans were to succeed.

Other experts were required, not only to act as independent witnesses to validate the find, but also as a form of academic reserve; 'tame scientists' who could be relied on to support the find, in case Woodward's backing dried up at any stage. On 20 April, Dawson visited his friend the Jesuit priest Marie-Joseph Pierre Teilhard de Chardin, then living in Hastings. Writing shortly after to his parents, Teilhard commented that the solicitor brought him some flint, animal bone and 'especially, a very thick, well-preserved human skull'. Other people to whom Dawson showed his prized 'discovery' were Samuel Allinson Woodhead, a chemistry instructor at Uckfield Agricultural College, Henry Sargent, a former curator of Bexhill Museum and Ernest Victor Clarke, a close personal friend.

Finally, on Friday 24 May, Dawson could wait no longer, especially now that Arthur Conan Doyle's *The Lost World* had already commenced serialisation in *The Strand Magazine*. He hopped on a train and travelled up to London. Walking into the Keeper of Geology's South Kensington office, Dawson placed the pieces of skull onto Woodward's desk, remarking 'How's that for Heidelberg?'. Woodward's reaction is, unfortunately, lost to us, but the find was evidently far greater than he could ever have expected. The cranial remains produced by Dawson were clearly of a human-like individual, stained a very dark brown by the gravels in which they had been entombed. It was their robustness that had clearly excited Dawson, the skull being abnormally thick, but without any trace of enlarged brow ridges. Could they really be part of an early ancestor to the human race?

Of course, we now know that the answer to this was an emphatic 'no'. In 1959, nearly fifty years after the Piltdown skull was first dramatically presented to Woodward in the quiet tranquillity of his London office, fragments of the cranium were subjected to radiocarbon dating. Radiocarbon (or carbon-14) dating is a method of scientifically assessing the age of something that was once alive. All living organisms contain carbon, a proportion of which is radioactive isotope carbon-14. When an animal or plant dies, it stops absorbing carbon-14 and the quantity present within the organism slowly starts to decay. As this rate of decay is theoretically known, the relative concentration of carbon-14 within a sample of bone or charcoal can be measured and an estimate of the date of death can be made. The Piltdown cranium was dated to 620±100 (GrN-2203), which may be calibrated at 95.4 per cent probability to having an origin somewhere between AD 1210 and 1480. The head of Piltdown Man, so vital in establishing the credentials of Britain's missing link was, at most, only 700 years old.

It is possible that the cranium, as well as being relatively young, had not actually travelled very far either. John Clements, writing to the magazine *Current Archaeology* in 1997, observed that the 'thickened cranium' may well have had its actual origins within the town of Hastings. In 1858, Mayor Alderman Thomas Ross was excavating on East Hill in the town, looking for a Roman lighthouse, but in reality finding an extensive early Medieval cemetery. At least one of the skulls, apparently part of a sixth-century grave deposit, was unusual being 'upwards of seven sixteenth's of an inch thick'. This, and other relics from the cemetery, remained in Ross' not inconsiderable private collection. Intriguingly, Dawson acted as solicitor for Ross and, following his death, seems to have inherited many of his finds. A reopening of Ross' trench 'back filled with many bones' occurred in 1902 during the building of East Cliff railway, itself a popular tourist attraction for many years after, whilst in 1912 drainage works conducted by the borough engineers near the cliff railway station 'uncovered two more unusually thick skulls'. None of this, of course, conclusively proves where the 'thickened' human cranium of Piltdown Man originated, but it does serve to demonstrate that an amateur antiquarian such as Dawson could, in the early years of the twentieth century, easily accumulate the basic skeletal remains necessary for the fraud.

Having established the nature of Dawson's find for himself in the security of his own London office, Woodward was keen to determine the provenance and exact circumstances of the find. This is the point at which Dawson's plan could have 'hit the buffers', for his story needed not only to be plausible but utterly watertight. Given the potential importance of his 'find', it needed to be verifiable in every last detail especially as it had never been in the gravels of Piltdown in the first place. Unfortunately, these are the elements that Dawson appears to have left incomplete at the time of his announcement of the great find to Woodward, leaving him to rely once again on a smokescreen of vague and ultimately unsupportable fiction.

One of the earliest accounts of the initial 'discovery' was provided by Dawson in the *Hastings and East Sussex Naturalist* for 1913. Presumably, this is the version of the story that he had prepared and provided for Woodward:

Many years ago, I think just at the end of the last century, business led me to Piltdown, which is situated on the Hastings Beds and some four or five miles north of the line where the last of the flint-bearing gravels were recorded to occur. It was a Court Baron of the Manor of Barkham at which I was presiding, and when business was over and the customary dinner to the tenants of the Manor was awaited, I went for a stroll on the road outside the Manor House. My attention was soon attracted by some iron-stained flints not usual in the district and reminding me of some Tertiary gravel I had seen in Kent. Being curious as to the use of the gravel in so remote a spot, I enquired at dinner of the chief tenant of the Manor where he obtained it. Having in remembrance the usually accepted views of geologists above mentioned, I was very much surprised when I was informed that the flint gravel was dug on the farm and that some men were then actually digging it to be put on the farm roads, that this had been going on so far as living memory extended, and that a former Lord of the Manor had the gravel dug and carried some miles north into the country for his coach drive at 'Searles'. I was glad to get the dinner over and visit the gravel pit, where, sure enough, two farm hands were at work digging in a shallow pit three or four feet deep, close to the house. The gravel is an old river-bed gravel chiefly composed of hand rolled Wealden iron-sandstone with occasional sub-angular flints. The men informed me that they had never noticed any fossils or bones in the gravel. As I surmised that any fossils found in the gravel would probably be interesting and might lead to fixing the date of the deposit, I specially charged the men to keep a look out.

Subsequently I made occasional visits, but found that the pit was only intermittently worked for a few weeks in the year, according to the requirements of the farm roads. On one of my visits, one of the labourers handed to me a small piece of a bone which I recognised as being a portion of human cranium (part of a left parietal) but beyond the fact that it was of immense thickness, there was little else of which to take notice. I at once made a long search, but could find nothing more, and I soon afterwards made a whole day's search in company with Mr A. Woodhead MSc, but the bed appeared to be unfossiliferous. There were many pieces of dark brown ironstone closely resembling the piece of skull, and the season being wet, any fossil would have been difficult to see. I still paid occasional visits to the pit, but it was not until several years later that, when having a look over the rain-washed spoil heaps, I lighted on a larger piece of the same skull which included a portion of the left supra-orbital border. Shortly afterwards I found a piece of hippopotamus tooth.

Dawson presented a more compressed version of this story to the Geological Society on the night of 18 December 1912, as part of his presentation, later published in the *Quarterly Journal of the Geological Society London*. In this account the formal dinner with the tenants of Barkham Manor is omitted, as is the 'whole day's search in company with Mr A. Woodhead'. The essential details of the find, namely the observation of flints in the road leading to the discovery of two labourers extracting gravel, remain the same, as does Dawson's request to the workmen to 'preserve anything that they might find'. One useful addition, however, is the approximate date of Dawson's discovery of a second 'larger piece of the same skull', which he cites as having occurred 'in the autumn of 1911'.

Although Dawson is vague as to exactly when he first encountered the gravel digging at Barkham Manor, other than to note it was probably 'just at the end of the last century', given that the discovery occurred during a meeting of the Court Baron of the Manor, it should be possible to supply a more accurate date. Dawson, in a letter to Woodward dated 22 May 1915, enclosed 'a notice of our four yearly meeting at Piltdown which led years ago to the discovery of the gravel bed'. As the meeting of the Court Baron to which Dawson draws Woodward's attention to was scheduled to occur on 'Thursday, the 10th day of June, 1915 at 12 noon', it must follow that there had been previous meetings in 1911, 1907, 1903 and 1899. Dawson had assumed stewardship in 1898, presiding over his first court on 27 July 1899. Such a date would fit perfectly with the solicitor's statement that he had investigated the gravel digging 'at the end of the last century', although Dawson's own (rather vague) observation to the Geological Society, that the first discovery of the gravel occurred 'several years ago' would seem to fit more with a meeting in either 1907 or 1904 (the four-yearly meeting scheduled for 1903 was, for some reason, postponed until 3 October 1904).

Unfortunately, attempts to pin down the date of Dawson's subsequent visit, where he was handed the 'small piece of a bone' by the farm labourers, does not appear in any official report. If Dawson was not tied to the visiting of the site during meetings of the Court Baron, which occurred on 3 October 1904, 10 May 1907 and 4 August 1911, he could have picked up the artefact at any time between 1899 and his 1911, when we know he found the second 'larger' fragment. A report appearing in *The Times* the day after Dawson and Woodward's presentation to the Geological Society was, however, rather more explicit concerning the date of the first discovery of skull, which it stated occurred 'four years ago'. If Dawson had been quoted correctly (inferring that the first piece of skull was handed to him in 1908), it seems odd that this rather crucial fact escaped mention in either the official or subsequent reports. Most writers on the Piltdown discoveries accept the suggestion that 1908 was indeed the year that Dawson was first handed a piece of skull.

Whatever the *exact* date that the first piece of skull was supposedly found, it is clear from Dawson's own account that there was a gap of several years before the larger piece preserving the 'left supra-orbital border' was finally located in 1911. What Arthur Woodward does not appear to have picked up on, either on receipt of Dawson's letter of 15 February 1912 (where he first informed him of the finding of the skull) or at any subsequent time (and especially after the presentation to the Geological Society in 1912), was the lengthy time gap between Dawson's first discovery and his reporting of the find. If the first skull fragment had indeed been found 'several years' before the second (and possibly as early as 1899), Woodward ought to have questioned why Dawson had not mentioned it to him before. Furthermore, as both Dawson and Woodward had in February 1912 only just examined a quarry near Hastings together (as set out in the opening paragraph of Dawson's letter of 15 February), why did Dawson not present or even discuss his find with Woodward then? Surely the earlier field trip would have been the perfect opportunity to show the artefact and gauge Woodward's opinion. A visit to Piltdown itself could even have been arranged in order to inspect the gravel at first hand. No explanation for this oversight on Woodward's behalf, nor indeed of Dawson's curiously long delay in reporting the find, is ever provided by either man.

Here we encounter another problem with Dawson's story, for if, as Dawson implies, the first meeting with the farm labourers excavating the gravel at Barkham Manor occurred in 1899, it follows that his request to them to 'keep a look out' for interesting finds did not bear fruit until around 1908. This point, left unremarked upon by Woodward, was picked by Joseph Weiner in 1955 who observed that 'the two labourers must have kept Dawson's request in mind, or been reminded of it, over a period of *some eight or nine years* before "one of the men" (and the context makes clear, one of the *same* men) at last alighted on a piece of cranium'. Is it really conceivable that Dawson's request was recalled by the workmen after such a lengthy time lag and with no artefacts recovered in the meantime?

In his book *The Earliest Englishman*, published after his death in 1948, Arthur Woodward presented a slight variant on the story of discovery. In this account, the workmen charged with extracting the gravel, after several unsuccessful visits to the area by Dawson:

> dug up what they thought was a coco-nut, and felt sure that this was the kind of thing which would please their curious and presumably generous friend. They could scarcely doubt that it was a coco-nut because it was rounded and brown and of the right thickness, with the inside marked in the usual way by branching lines and grooves. It seemed a familiar and common object, but, as it was a little bulky to keep, they broke it with a shovel and threw away all but one piece, which they put in a waistcoat pocket to show Mr Dawson on the first opportunity. When he came round again, the men produced their find and described to him the 'coco-nut' from which they had broken it. They showed him the place where they had found it, and told him that the pieces which they had thrown away were in the heaps of rubbish around. Mr Dawson recognised at once that the supposed coco-nut was really a human skull of unusual thickness and texture, which had been hardened and stained brown by oxide of iron in the gravel. He did not show any excitement or concern about the misfortune that had happened to the unique fossil, but he patiently waited for a favourable occasion to examine the pit and see whether the labourers' story was true.
>
> Time after time Mr Dawson visited the spot and searched the rain washed heaps of gravel, but it was not until a few years had elapsed that he found a second piece of the skull, which fitted exactly one broken edge of the fragment which the men had given to him. Renewed search eventually unearthed a third piece which fitted the other two, and then came two more separated pieces which certainly belonged to the same skull. The men's story was thus confirmed, and it was evidently desirable to dig up and sift all the gravel which remained in and around the pit.

The origins of this particular version of the story are unclear. Dawson's original statements describe the primary find as a single 'small piece of a bone'. Nowhere does Dawson himself relate a story, at least in print, recounting the unearthing of a more complete skull that was deliberately smashed in the misguided belief that it was a coconut, and yet Woodward appears quite sure that this was how it actually happened. The origins of the 'coco-nut' story can only have lain with Dawson himself, for no one ever claims to have interviewed the two anonymous farm hands who could have verified the statement, and Woodward does not appear to have spoken in detail with any of Dawson's early colleagues, such as Samuel Allinson Woodhead.

It is possible that the incident was concocted by Dawson as a way of lending colour to the story of how the piece was first found, but that it was only 'conversational evidence', never intending to be put down onto the printed page. This may well be true, for certainly William Lewis Abbott, jeweller and amateur geologist, related a similar story in February 1913, whilst *The Times*, telling the details of Dawson's presentation to the Geological Society in December 1912, commented that one of the labourers originally gave Dawson 'a fragment of a human skull which they had just discovered and had evidently broken up and thrown away'. No mention of a coconut but a similar idea, albeit different from the account provided by Dawson in the official reports of 1913.

The version of the story published in the *Quarterly Journal of the Geological Society* for that year does suggest why so many cranial fragments of the same individual had been found scattered across various spoil heaps: 'apparently the whole or greater portion of the human skull had been shattered by the workmen, who had thrown away the pieces unnoticed.' But this is not the same as the deliberate breaking up and discarding of a 'coco-nut', for here the labourers were not aware of their actions. Woodward does not appear to question the obvious divergence in accounts, and neither does he comment upon the strange idea that two labourers, having specifically been asked to look out for unusual artefacts, managed to deliberately destroy the *only interesting thing* that they had seen in eight years of digging.

There is one final, rather important problem with the various accounts supplied by both Dawson and Woodward that explain the initial discovery of Piltdown Man: namely the supposed location of the gravel extraction pit. Dawson says that, having seen 'iron-stained flints' in the road outside the manor house, he was surprised to hear, that not only were the flints derived from the farm, but that 'some men *were then actually digging it*'. Why should the location of the extraction pit and the fact that two men were working it have surprised Dawson? In order to have gained access to Barkham Manor so as to attend the meeting, Dawson would have passed directly in front of the gravel pit; he simply could not have missed it. Why, then, given his recorded interest in the sort of artefacts accidentally unearthed by labourers, did he not pause on his way in to the meeting to inquire of the labourers what, if anything, they had found? Why did he not remember the pit when told by the chief tenant of the manor where the unusual flints had been obtained? Once again, Woodward does not question the obvious discrepancies in Dawson's account, nor wonder why the published version describing the location of the pit was so obviously misleading.

Somehow, Dawson's talent at creating fiction had survived intact. The Piltdown skull was evidently considered to be too important by Woodward and others in the scientific establishment to be brought down by any inconsistencies in the story of how it had been discovered, when and by whom. Those who knew Dawson, and had good experience of his 'amazing' series of finds, however, would have experienced a clear sense of déjà vu. The tale of anonymous labourers finding something incredible which, thanks to Dawson's timely intervention, is saved but then not reported on by the solicitor for years was one that was all too familiar, having an uncanny resemblance to the circumstances surrounding the Beauport Park statuette, Blackmore's stone axe, the Bexhill boat, the Uckfield horseshoe, the Pevensey brick and the Toad in a Hole.

The first objective piece of evidence concerning the existence of the Piltdown skull is Dawson's letter to Arthur Woodward dated 14 February 1912. Before this, as far as it may be ascertained, Dawson confided his 'discovery' to no other person, not even his wife Helene. As Dawson was the mastermind behind *Eoanthropus*, we cannot logically accept anything that he tells us concerning the find prior to 14 February 1912 as being in any way truthful. The date of 1908 provides for Piltdown, just as it did for Pevensey, Bexhill and Beauport Park, a significant delay between the recovery of an artefact and its eventual reporting (nearly five years for the Pevensey brick, five years for the Bexhill boat and a full ten years for the Beauport statuette). Such a lapse provides, once more, a smoke-screen obscuring the precise circumstances of discovery. Anyone trying, in 1912, to record the *exact* details of when, where and how the first pieces of skull were found, would have to rely solely upon Dawson's testimony, the original anonymous 'workmen' of the story having long since disappeared.

As the first confirmation by Dawson of the existence of human remains at Piltdown occurred as late as February 1912, it is apparent that, against the sequence of events established in most accounts of the hoax, that Doyle's novel, *The Lost World*, actually claims precedence over Dawson's find. Clearly, in writing the book, Doyle was inspired in part by his own surroundings, with the many fossils – including his own discoveries – that had been made in the Sussex Weald, and by those whom he had recently corresponded (for example Professor Summerlee in *The Lost World* has a more than passing resemblance to Arthur Woodward). The savage 'ape-men' in the book, however, possess few of the traits that were later to be associated with Piltdown Man; Doyle's lost race have ape-like skulls (Piltdown's skull is human), 'thick and heavy brows' and, rather crucially, 'curved, sharp canine teeth' (quite unlike those later recovered from the gravels of Barkham Manor). If anything, the apparent coincidence between the timing of *The Lost World* and the discovery of Piltdown Man is nothing of the sort; the idea for the hoax coming directly from the pre-publication manuscript of Doyle's novel, and not the other way around. Doyle was the unwitting inspiration for Dawson's 'missing link'. All Dawson had to do, if he were to successfully remove suspicion that the great author's novel had pre-eminence over his find, was to push the initial discovery of his primitive man back to 1908.

Arthur Woodward's analysis of the first skull fragments left him in no doubt as to the potential significance of the find. Woodward was later to recall that, in the discussion that followed the preliminary examination, he and Dawson decided to commence work in the gravel pit almost immediately in the hope of retrieving more remains. Dawson swiftly obtained permission for work from the landowner, George Maryon-Wilson and the tenant farmer, Robert Kenward, both of whom he knew well, then recruited his old fossil-hunting companion Father Marie-Joseph Pierre Teilhard de Chardin.

Work began on 2 June 1912, the greater part of the day being lost to the preparation of a picnic. By the time Dawson, Teilhard and Woodward arrived at Barkham Manor gravel pit, it was already the afternoon. Fortunately, Teilhard recalled, 'a man was there to help us dig'. Although not specifically mentioned at the time, it is probable that this workman was Venus Hargreaves, who appears to have been the sole labourer employed throughout 1912 and 1913. The strategy employed during the first few days would appear simple enough: Dawson, Teilhard and Woodward were to inspect all existing spoil heaps surrounding the area of recent gravel extraction, whilst

Hargreaves generated more spoil through the enlargement of the original pit. This arrangement suited Dawson well, for it meant that any further discoveries relating to Piltdown Man need not be planted in the ground, they only needed to be secreted within the looser soil of the spoil heaps. The gathering together of Hargreaves, Woodward and Teilhard (whom Dawson had fooled before at Hastings quarry) meant, furthermore, that the solicitor would not have to make all the 'discoveries' himself: now he had others who could find material and divert suspicion away from himself.

The work was at first painstakingly slow, Dawson later noting that 'every spade-full had to be carefully sifted and examined' so as not to miss even the smallest of artefacts. Agonising though this must have been, especially for Dawson who had been keen to examine the pit for some considerable time, Woodward observed that it was an advantage 'because the discovery of bones and teeth, all stained brown, in a dark-coloured gravel, which was full of bits of ironstone and brown flints, needed very close and slow examination of every fragment'. Employing more than one labourer, though at face value a time-saving initiative, would not, Woodward noted, have been altogether satisfactory 'because every spade full had to be watched, and generally passed through a sieve'.

The excavations continued sporadically throughout June, July and August of 1912, and, in the absence of specific dates, any of the finds generated could conceivably have been recovered at any point within this broad time frame. Unfortunately, neither Woodward nor Dawson appears to have kept a notebook in order to chronicle the events as they unfolded (or if they did, such a work has not survived). In fact, the only time that a key date may plausibly be cited for the 1912/13 season is when a visitor to the excavation, such as Teilhard, made note of it in their own diary or personal correspondence. The closest that Arthur Woodward himself came to recording the sporadic nature of the gravel pit examination is when he observed, in his later book *The Earliest Englishman*, that 'we were both well occupied with ordinary duties during the week, so we could devote only our weekends and occasional holidays to the task'.

Dawson's summary account for the pages of the Geological Society's *Journal* noted the discovery of unspecified numbers of cranial fragments from the spoil heaps, the 'right half of a human mandible' and a 'small portion of the occipital bone' from the undisturbed gravel, teeth of elephant, mastodon, beaver, horse and hippopotamus, as well as a red deer antler, a deer metatarsal and a number of worked flints. The implication is that these were found throughout the 1912 season. Woodward's later account, in *The Earliest Englishman*, is unfortunately no real help, as it conflates all elements of the fieldwork into a few short sentences. Perhaps, by the time Woodward came to compile the book (which, although not published until 1948, appears to have been at least started no later than 1915) he had become unclear as to the *exact* order that material had been found in. Re-reading Dawson's comments on the 1912 dig cannot have helped.

In the account of the excavation supplied to the *Hastings and East Sussex Naturalist* in 1913, Dawson observed that the first stage of investigation at Barkham Manor did not go terribly well for 'there were many days of most unpromising work. It was not until we had been busy off and on for some weeks that after a hard and unproductive day's work I struck part of the lowest stratum of the gravel with my pick, and out flew a portion of the lower jaw from the iron-bound gravel.' Teilhard, in a letter to his parents on 3 June 1912, however, commented that, despite not having arrived on site for the first day's digging until after 3 p.m., the team 'worked for several hours and finally had success. Dawson discovered a new fragment of the famous human skull; he already had three pieces of it, and I myself put a hand on a fragment of elephant's molar.'

The crucial aspect here is that Teilhard recalled the events of the first day not less than *twenty-four hours after they had occurred*. The initial phase of work at Barkham Manor had, therefore, proved to be highly successful, with strong indications that, not only were more pieces of the human skull in evidence, but that more was undoubtedly to come. Quite why Dawson (and much later Woodward) provided such a downbeat account of the excavation's 'unpromising' beginning, remains a mystery. Perhaps neither man could remember the exact order in which things were discovered, or perhaps both were embarrassed that spectacular finds were retrieved within a very short time of work having got underway.

As the work progressed over the following weeks, more exciting discoveries came to light, Woodward noting later that 'in one heap of soft material rejected by the workmen we found three pieces of the right parietal bone of the human skull – one piece on each of three successive days. These fragments fitted together perfectly, and so had evidently not been disturbed since they were thrown away.' At some point in late June (possibly Saturday the 23rd), Dawson found the crucial fragment of skeleton which could at last be used to compare the Piltdown specimen with that of *Homo heidelbergensis*: a jaw. The jaw derived from what Dawson and Woodward were later to describe as 'a somewhat deeper depression of the undisturbed gravel', Dawson himself observing that 'so far as I could judge, guiding myself by the position of a tree 3 or 4 yards away, the spot was identical with that upon which the men were at work when the first portion of the cranium was found several years ago'.

The jaw was an important piece of evidence for it appeared to have derived from a patch of *undisturbed* gravel, therefore making it the first in situ discovery from the entire investigation. It demonstrated that, not only was this indeed the correct place to dig (and that the earlier cranial fragments had not merely been introduced from somewhere else) but also that the activities of the earlier farm labourers had not completely eradicated all trace of their early ancestor. In an alternative account of the dig, Dawson relates the circumstances of his priceless discovery: 'I struck part of the lowest stratum of the gravel with my pick, and out flew a portion of the lower jaw from the iron-bound gravel.'

Although vital for the purposes of the fraud, the size of the mandible created a specific problem for the Piltdown forger. The piece was too bulky an artefact to be found, casually discarded, on the spoil heap, as all soil was being carefully examined prior to removal, and it was too precious a piece to be retrieved at the end of a pick or shovel. This created a unique set of issues, for in order to simulate plausible ground conditions, Dawson needed to fool all members of the excavation team into thinking they were digging into virgin soil. Here the forger was presented with one of two possible solutions to the quandary. First, he could bury the piece (difficult in itself without attracting attention) and get someone unfamiliar with the complexities of the gravel to commence excavation over the area. Certainly there were many inexperienced volunteer helpers visiting the excavations during the course of the 1912 season, but would any of them really have been entrusted with the removal of soil in a sensitive and untested part of the trench?

The second option available to the forger with regard to the recovery of the ape-like jaw was to actually 'find' the piece himself. To do so may, quite naturally, attract some suspicion, especially if he had already recovered a great many other finds from the site, but the advantage was that the hoaxer would not actually have to plant the find in the gravel or in some way disguise its presence there. All Dawson had to do was to convincingly provide the impression that it indeed been in situ and, to this end, an independent (and unwitting) witness would therefore be required. Thus the circumstances of the 'discovery', dislodged by Dawson's vigorous picking at the supposedly undisturbed gravel bed, can be explained, Woodward later noting that the piece 'had evidently been missed by the workmen because the little patch of gravel in which it occurred was covered with water at the time of year when they reached it'.

The mandible 'found' in 1912 was an absolutely crucial find demonstrating that, although the cranium of Piltdown Man undoubtedly possessed human characteristics, the face was very ape-like; missing a chin and with a prominent bony thickening, known as the simian shelf, along the front. Unfortunately, for those wanting to definitively prove that jaw and skull belonged to the same individual, the new find represented less than half of the original jaw: only the right side survived, and it was missing its articular condyle, the point of contact between jaw and cranium. This was, of course, deliberate for, had this point of articulation survived, it would have been abundantly clear that the two elements recovered from the gravel pit were from two very different species.

The absence of the condyle meant that attention focussed upon the two surviving molars, which had a wear pattern unlike that encountered among apes. Woodward was later to note the flattened crowns and 'non-aligned' surfaces of these two in situ teeth, which suggested that Piltdown Man had possessed a rotary chewing action akin to modern humans. The examination of the teeth under a microscope in the 1950s, however, revealed very fine scratch marks demonstrating the presence of an abrasive. Just like the teeth of *Plagiaulax dawsoni*, the shape of Piltdown Man's teeth had not been the result of natural wear, but the application of sandpaper and a metal file.

When, in 1959, the orang-utan mandible was subjected to radiocarbon dating, it produced a date of 500±100 (GrN-2204), which may be calibrated at 95.4 per cent probability to between AD 1290 and 1640. These results suggested that, though the skull appeared to be less than 800 years old, the jaw was, in all probability, considerably younger. A second sample from the jaw was processed in the late 1980s, following advances in dating procedure, providing a determination of 90±120 (OxA-1395), which may be calibrated at 95.4 per cent probability to between AD 1630 and 1960. The obvious discrepancy between the two dates provided for the same mandible has never been satisfactorily explained, though the new determination would, the investigation team noted, 'accommodate a postulated 19th century date' for the piece. It may well have been relatively 'fresh' at the time it was inserted by the forger into the gravels of Barkham Manor.

As concerns the means to acquire an orang-utan jaw, Joseph Weiner, chief investigator of the fraud in 1953, observed that a man such as Dawson would have had no difficulty in procuring unusual items, such things could easily having been 'bought from, or through, a local taxidermist, or, if not, then easily enough from one of the famous London firms'. One such firm, Weiner noted, was 'Gerrard's in Camden Town' where 'unmatched jaws and other odd bones' were relatively easy to procure in the years before the First World War. There was also the monthly taxidermists' auction at Stevens in King Street, where 'a collector could hope to pick up an enormous variety of specimens and bones'. Various skeletal pieces 'such as teeth and man-dibles' could, Weiner observed, be purchased from the auction 'and many geologists had them'.

But did Dawson acquire the pieces necessary for the fraud from such a place? There is no record of him ever having collected 'ape bones' at any time in his archaeological or palaeontological career, but then if he had been the forger he would certainly have covered his tracks. As the prime mover behind the Piltdown hoax, Dawson would have found many of the basic building blocks necessary to accomplish the forgery read-ily to hand in Sussex. As the co-founder of the Hastings and St Leonards Museum Association, and also a member of the museum committee in charge of the acquisition of artefacts, Dawson would have seen a vast array of material. Whilst performing his legal role, the solicitor also acted on behalf of a number of prominent antiquarian collectors, cataloguing mate-rials bequeathed to the museum. One such assemblage was the Brassey Collection, comprising a diverse set of materials brought back to England by Lord and Lady Brassey after their extensive round the world sailing tours. Interestingly, the Brassey's yacht, *The Sunbeam*, visited Malta, Crete and Sarawak and North Borneo, places which have been chemically iden-tified as the provenance of both the mandible and the associated Piltdown animal assemblage.

Having acquired the jaw, selectively removed the diagnostic elements, detached all but two of the teeth and filed the remainder down in order to produce a more 'human' wear pattern, Dawson needed one last touch: the mandible had to be the same colour as the cranium. In order to create the impression that the bone pieces had lain in the iron-rich gravel beds of Barkham Manor for millennia, Dawson stained both a dark 'chocolate' brown by dipping them in a solution of potassium bichromate.

That Woodward was aware of Dawson's treatment of the Piltdown bone assemblage is evident from his later comments in *The Earliest Englishman* where he notes that the skull bones had all been 'hardened and stained brown by oxide of iron'. 'The colour of the pieces,' he went on, 'which were first discovered was altered a little by Mr. Dawson when he dipped them in a solution of bi-chromate of potash in the mistaken idea that this would harden them.' Woodward apparently accepted Dawson's reasons for conducting the treatment, to better preserve the bone, though he observed that it did not in any way affect the hardness of the bone, merely that it altered the surface colour 'to a minor degree'. What he never questioned, however, was why Dawson had continued treating the bone, even after the relative failure of the hardening experiments, nor why he subjected certain pieces of the Barkham Manor flint assemblage to the same treatment. Flint is a stone which, in hardness and durability, is second only to diamond. It does not require any form of artificial hardening in order to ensure long-term preservation, but it does require staining if it is to look as if it has derived from the same source as the bone.

If the application of bichromate solution by Dawson to the Piltdown bone had been entirely innocent, we are faced with a rather interesting, not to say worrying, observation. The first finds of bone at Barkham Manor were passed on to Woodward only after an unspecified, but presumably rather lengthy, period of time. Whilst in his possession, Dawson evidently treated these artefacts in order to harden them. The artefacts retrieved during the course of the 1912–14 excavations, such as the mandible and canine, were, however, not entrusted to Dawson's care, but were taken almost immediately back by Woodward to the Natural History Museum in London. The fact that later discoveries (such as the jaw) had also been chemically treated in the same way means that either Woodward or the technicians at the museum themselves used bichromate solution (which is unlikely, given Woodward's understanding that it did not affect the hardness of the samples, only their colour), or that they were chemically treated *before being planted* at Barkham Manor.

What, then, about the facilities necessary to perpetrate the Piltdown hoax? Where did Dawson create his fraud? A clue may be given by various contacts who approached the Natural History Museum following the initial exposure of the hoax in the mid-1950s. One such letter, emphatically linking Charles Dawson to the forgery, was received by the Keeper of Geology from a Captain Guy St Barbe. Joseph Weiner and his assistant,

Geoffrey Harrison, duly conducted a formal interview with St Barbe at his St Alban's home on 4 January 1954. Key in recollections of St Barbe were a series of unannounced visits to Dawson's Uckfield office. One time, St Barbe appears to have surprised Dawson whilst he was actually in the process of staining bones. 'There were,' St Barbe recalled, 'perhaps 20 dishes, 6 inches in diameter, full of brownish liquid.' Dawson, when challenged, explained that he was simply 'interested in staining bones'.

Twenty three days later, on 27 January, St Barbe was again interviewed, this time by Kenneth Oakley, and provided slightly more detail concerning the incidents in Dawson's Uckfield office. 'Dawson was,' Oakley was to record, 'surprised in his office on three occasions. First by Barbe who noted a strong smell as of a chemist shop and several dishes containing dark-coloured fluid, in some of which bones were being boiled … Dawson was very agitated and explained that he was experimenting with a view to finding out how bones and flints become stained.' On the timing of these 'three occasions', St Barbe was imprecise, though he felt that 1910 was the earliest possible year during which this could have happened.

St Barbe could not provide exact details of what sort of bone was being stained, nor whether these pieces had certainly derived from Piltdown, precise details being difficult to remember after the passing of some forty years. Of course, the intimation was that here Dawson had been 'caught in the act' of forgery. If Dawson had been using his own solicitor's office as a form of secret laboratory, it would appear strange that he did not possess the wit to secure the door, but then would he have really expected a sudden, unannounced visit of the sort perpetrated by St Barbe? Who is to say that the temptation to work on scientific experiments in his office, away from the pressures and demands of home, was not an irresistible one? In any case, we have, by Dawson's own admission, details concerning just such extracurricular activities.

On 3 July, for instance, shortly after recovering a piece of human cranium from a new site at Barcombe Mills (see below), Dawson wrote to Woodward asking if he could obtain a 'recipe for gelatinising, as the bone looks in a bad way and may go wrong in drying'. Intriguingly, he added 'I have got a saucepan and gas stove at Uckfield', a comment that appears to support St Barbe's observations. Quite when Dawson found the time to pursue legal matters is never made clear.

Charles Dawson was not the only one to find human remains during the initial season of 1912, as the later report on the excavations noted:

'Dr. Woodward also dug up a small portion of the occipital bone of the skull from within a yard of the point where the jaw was discovered, and at precisely the same level'.

Contrary to the account given by Dawson in this part of the article appearing in the *Quarterly journal of the Geological Society*, however, the small portion of occipital bone was not certainly dug up from undisturbed gravel, but from a nearby spoil heap, a point made by Woodward in his later book: 'I found *in another heap* an important fragment ... which fitted the broken edge of the occipital bone and gave us the line to contact with the left parietal bone.'

The piece in question had therefore *already been removed* from its original context before Woodward discovered it. It is possible that Dawson had meant to say that his colleague had 'dug the bone up from the top of the spoil heap', a relatively minor mistake in phrasing perhaps, but a rather crucial difference from finding the piece in situ.

A small but diverse collection of faunal remains was assembled during the 1912 investigation at Barkham Manor. In their report to the Geological Society, Dawson and Woodward recorded that they had retrieved 'two small broken pieces of a molar tooth of a rather early Pliocene type of elephant, a much-rolled of a molar of *Mastodon*, portions of two teeth of *Hippopotamus*, and two molar teeth of a Pleistocene beaver'. On the surface of the field to the immediate west of the hedge flanking the gravel pit, the team furthermore found 'portions of a red deer's antler and the tooth of a Pleistocene horse'. As to how these important artefacts had come to be lying on the surface of the field (rather than in the more deeply bedded gravels), Woodward theorised that they may have been 'thrown away by the workmen' or possibly churned up by more recent ploughing activity. The reality was, of course, that Dawson had placed the pieces in the ploughsoil, awaiting their recovery by the team.

Apart from the bone assemblage, a number of worked flints were also recovered during the first season. Of course, it had been the flint in the trackways around Barkham Manor that had apparently drawn Dawson to the gravel pit in the first place. The discovery of humanly worked, or 'knapped', flint at Piltdown would prove a significant boost to the excavations, demonstrating the presence of early prehistoric activity. It was even possible that the flint tools themselves could have been generated by the ape-like human the team was in the process of investigating. That would, in turn, suggest an intelligent individual being capable of rational thought.

The first flints recovered from Piltdown were generally dark brown, similar to the artificially stained jaw and cranium. They comprised 'several undoubted flint-implements, 9 besides numerous "Eoliths"'. 'During the late nineteenth and early twentieth century 'eoliths', or 'stones of the dawn', were a popular subject for sometimes quite heated debate. These crudely broken flints were considered by some to represent the first human tools, though others viewed them less charitably as the result of natural breakages. The main problem with these 'dawn stones' was that, unlike the beautifully worked flint axes, scrapers, projectile points and digging tools of later prehistoric periods, the typical eolith often comprised no more than a simple sharp edge, sometimes with signs of battering. Supporters of the eolith argued that frost shattering or river bed tumble could not produce the one-sided chipping evident on a number of specimens. The apparent association of eoliths with the bones of extinct mammals and the possible traces of early human activity appeared to increase the likelihood that these stones had indeed been deliberately created.

Dawson was well aware of the controversy that then surrounded eoliths, noting that 'if these specimens are subsequently proved to be "artefacts" they must belong to the early dawn of the formation of implements by man'. If the eoliths were to be accepted as genuine products of early human endeavour, then it was vital that an accurate provenance be obtained. Unfortunately, the solicitor was vague as to exactly *where* the material had originated, noting only that it occurred 'both in the gravel-bed and on the surface of the plough-lands'. Later on he was to discount the eoliths completely as artefacts, quite probably because they had not been part of his master plan. Rather than being designed and created by the hoaxer in order to show the range of Piltdown Man's capabilities, it would seem that the eoliths recovered from Barkham Manor really were just naturally broken flints: pieces that happened to be located within the gravel. Certainly the brown surface of these 'dawn flints' did not result from artificial staining processes and, unlike the 'palaeoliths', they were not affected in any way by the application of hydrochloric acid. Dawson had been largely sceptical of the artefacts from the start of the investigation, something which may further indicate that he never intended them to be part of the hoax.

With regard to the rest of the flint assemblage, only four or five 'palaeoliths' were ever produced by Dawson in evidence for the workmanship of Piltdown Man. Any fieldworker worth his salt could gather such material up from the ploughed fields of southern England, and, in the early years of the late nineteenth and early twentieth centuries, many private antiquarian lithic assemblages were known to exist. Dawson himself had no mean collection of worked flint, including serrated blades, spears and polished axes and flints from Denmark, flakes from Birling Gap near Eastbourne, and various arrowheads from North America, some of which were passed on to Hastings Museum after his death. A complete listing of the Dawson lithics, as held by Hastings Museum, interestingly shows that Dawson also possessed four modern copies of implements, made by Flint Jack, a famous nineteenth-century fabricator of prehistoric-looking tools.

The Barkham Manor lithics, though repeatedly referred to as palaeoliths, inferring an early date for their use, are more likely to represent unfinished stages in Neolithic flint axe manufacture. Many flint mines are known to have existed along the chalk ridge of the South Downs during Neolithic times, concentrating in areas such as Cissbury, Harrow Hill and Blackpatch Hill behind modern-day Worthing. A walk across these hills today will reveal large numbers of struck flints, in rabbit burrows, molehills and plough ruts, the by-product of prehistoric mining and flint-shaping activities. When these sites were first scientifically examined, in the mid-nineteenth to early twentieth century, many 'roughouts', the unfinished residue of Neolithic axe manufacture, were actually believed to represent finished examples of Palaeolithic workmanship. The association of Palaeolithic tools with obvious signs of deep mining led a number of prominent scholars to suggest that flint mines were not Neolithic, but were thousands of years older, dating to the earliest phases of human occupation in the British Isles. The Barkham Manor flint certainly fits more comfortably within a Neolithic lithic assemblage than anything yet recorded from the British Palaeolithic. The flints' artificially stained surfaces, apparently conducted in order to disguise their origin (from the white chalk not the brown gravels of Piltdown), further increases the likelihood that they represent Neolithic artefacts, not the product of early human tool manufacture.

The staining of the Barkham palaeoliths is only superficial; their white, chalky patina having been effectively disguised by an application of bichromate solution of the sort that we know Dawson used to 'harden' the Piltdown bone. Of course, the lithics did not require artificial hardening, and so the use of chromium here can only have been in order to disguise the surface colour of the material, as flints with a white cortex do not naturally appear in the gravels of Piltdown. Dawson is, as far as can be ascertained, the only person to have used bichromate solution on the Piltdown artefacts, Woodward commenting on the fact on a number of occasions.

Within weeks of the completion of the fieldwork in August 1912, rumours were starting to circulate about the startling nature of the Piltdown discoveries. On Thursday 21 November the *Manchester Guardian* printed a story leaked by an anonymous informant. Under the headline 'The Earliest Man? Remarkable Discovery in Sussex', the paper observed that 'One of the most important prehistoric finds of our time has been made'. Woodward was unimpressed. Full disclosure of the Piltdown remains was not due to take place for another month, giving himself and Dawson more time to prepare their statements and complete the first reconstruction. His exasperation was increased when a number of reporters descended upon his office at the Natural History Museum and demanded to know whether the story was true. Woodward confirmed the story, but provided only circumstantial detail, asking the journalists to be patient.

Dawson pretended to be similarly exasperated that the press had got hold of the story, but the likelihood is, given his association with the media on earlier discoveries, he realized the value of the press in heightening public awareness; that it was he who had, in fact, been the anonymous informant. Certainly, press speculation concerning the find helped excite both public and academic interest. As a consequence, the numbers descending upon London's Burlington House, where the meeting of the Geological Society was to take place, were unprecedented, one reporter observing that 'never has the meeting room been so crowded'. The presentation began at 8 p.m., with Dawson providing a summary of the events leading up to the initial discovery, the geological background and the preliminary results of the first season's excavation.

Woodward's presentation focussed upon the interpretation of the Piltdown skull and the technical nature of its reconstruction. There could be no doubt, he stated, concerning the perceived humanity of the skeletal remains. In conclusion, he noted that the characteristics of the Piltdown skull presented the question of 'whether it shall be referred to a new species of *Homo* itself, or whether it shall be considered as indicating a hitherto unknown genus'. Given that the surviving facial aspects of the skull appeared to differ so much from modern humans, he proposed that 'the Piltdown specimen be regarded as the type of a new genus of the family Hominidae … named *Eoanthropus dawsoni*, in honour of its discoverer'. Dawson's 'Man of the Dawn' had arrived.

In the lively discussion that followed the presentations, most seemed to agree with the conclusions made by the excavation team, but a number took exception to the proposed age of the early human. Dawson and Woodward had made the case that the stratified gravel at Piltdown was 'of Pleistocene age', but noted, as a word of caution, that its lower levels contained animal bone 'derived from some destroyed Pliocene deposit' close by. 'Pliocene' and 'Pleistocene' are geological terms which, at the time the Piltdown remains were discovered in the early years of the twentieth century, possessed relatively elastic qualities. Today we interpret the Pliocene as the fifth epoch of the Cenozoic (or era of 'recent life'). The Pliocene, which spanned the period between 5.2 million and 1.6 million years ago (being preceded by the Palaeocene, Eocene, Oligocene and Miocene), was an epoch of geological change in which the continent of modern-day Africa collided into Europe, creating the Mediterranean, and South America impacted with the North. The Pleistocene, which followed the Pliocene, is often referred to the 'Great Ice Age'. In fact, it was a time of at least nine major periods of glacial advance, when ice sheets covered anything up to between 20 and 30 per cent of the world's surface.

At the maximum point of glacial advance, during a period known as the Anglian (which began approximately 478,000 years ago and ended around 423,000), the ice sheet over Britain extended as far south as northern Cornwall, following a line towards modern-day Bristol, north London and out towards Essex. The Holocene, or modern epoch, began at the end of the last ice age around 10,000 years ago. Quite when, where and how the earliest humans fit into this, remain popular topics of debate to this day. The present consensus is that early humans, or '*hominids*', first entered the north-western European peninsula around half a million years ago, before the Anglian glacial advance. The 'British Isles' simply did not exist at this time, the area of modern-day Britain being joined to the rest of continental Europe (the waters that separate them today being locked up in an immense ice sheet covering the North Sea).

In 1912, although the basic processes of geologic change were understood, the antiquity of human endeavour was most certainly not. The question of where *Eoanthropus dawsoni* fitted within 'earth time' was therefore crucial if Charles Darwin's evolutionary model was to be fully understood. Dawson and Woodward suggested that their 'Man of the Dawn' had probably existed in an early phase of the Pleistocene (probably 'during a warm cycle'), an epoch that was then estimated to be 'certainly one hundred thousand years ago, perhaps as much as five times this sum'.

Following the furore of the pre-Christmas meeting at the Geological Society, Dawson and Woodward returned to their normal lives; Woodward to the Natural History Museum and Dawson to his practice in Uckfield. The first introduction of *Eoanthropus dawsoni* to the world had been a triumph; now Dawson had to capitalise on this success. He needed to make sure that not only were there more finds for the excavation team to discover in the seasons that followed, but also that these new discoveries specifically resolved the many questions and queries of the Piltdown doubters: critics who believed that the remains so far recovered from Barkham Manor were not those of the much sought after 'missing link'.

Sometime in late May 1913, Woodward and Dawson recommenced their excavations of the Piltdown gravel. As in the previous season, much of the heavy earth-moving activity was conducted by Venus Hargreaves, whilst Dawson and Woodward sifted through the spoil. Excavation and examination of the gravel was conducted on a methodical, if painfully slow, basis; excavating, sieving and minutely sorting soil derived from the pit. On the weekend of 9–10 August, the team were joined by Teilhard de Chardin, just recently returned from France, and Maude Woodward, Arthur Woodward's wife. It was on the Saturday, a point confirmed in a letter from Teilhard to his parents, that a small but important discovery was recorded by Dawson: two nasal bones lying '2 or 3 feet from the spot where the mandible was found'.

The nasal bones were, perhaps unsurprisingly, a match for the speculative recreation in plaster that Woodward had created at the end of 1912; they were *exactly* what the team had been hoping to find in order to support the reconstruction. Dawson would, no doubt, have preferred it if somebody else within the team had found the pieces, thus diverting any suspicion of fraudulent activity away from himself. With Teilhard and Mrs Woodward accompanying the excavators that Saturday, he certainly had a better chance of ensuring that another member of the expedition made the big 'discovery' but perhaps, given the importance of the pieces, combined with their small and fragile nature, he felt it best on the day to make the find himself, albeit with witnesses around him. That way he could ensure that not only were both fragments successfully recovered, but also the exact circumstances of context could safely be obscured and destroyed in their retrieval.

When the remains of Piltdown Man were critically assessed by Joseph Weiner in the early 1950s, it was discovered that the 'nasal bones' were not in fact human at all, nor originally part of a nose; both pieces having been carved from the 'limb-bone' of an unspecified animal. Just like the cranium, the bone had been artificially stained in order to give the impression that they had lain in the iron-rich gravels of Barkham Manor for millennia. Dawson, it would appear, had once again been busy with his whittling knife.

Despite the 'finding' of skull and 'nasal' bones, the excavation team were no closer to ascertaining whether Dawson's Man of the Dawn had an ape-like or humanoid face. There had, in this regard, been significant debate in academic circles since the previous December as to the nature of the missing canines in the damaged jaw of *Eoanthropus*. Canines are the longer, stronger, pointed teeth in any given jaw which evolved, in all probability, to hold food and help tear it apart. Mammals possess four such teeth, two in the upper maxillary (popularly referred to as 'eye teeth') and two in the lower man-dibular arch. Ape canines are larger than those in humans, and the arguments surrounding the Piltdown remains had, by the start of 1913, centred around whether the absent teeth had been elongated, as in modern apes, or whether they had been smaller and more human in form. Woodward had preferred to visualise the canines of *Eoanthropus* as being approximately halfway in size between those of ape and man, and these were the ones that he had favoured for his plaster-cast reconstruction of the skull.

In order to resolve the matter, the Piltdown excavation team really needed to find more pieces from the fragmentary jaw; preferably more teeth. What happened next, then, was extremely fortuitous. 'We contin-ued to work,' Woodward was later to recall, 'without much success until Saturday, August 30th.' Having methodically processed the excavated gravel, Woodward and Dawson observed that Teilhard, then excavating a particu-larly deep area without Venus Hargreaves, who was absent that particular day, looked exhausted and they suggested that he take a break from hard labour and take a 'comparative rest' by further sorting some of the rain-washed gravel. 'Very soon,' Woodward went on, 'he exclaimed that he had picked up the missing canine tooth, but we were incredulous, and told him we had already seen several bits of ironstone, which looked like teeth, on the spot where he stood. He insisted, however, that he was not deceived, so we both left our digging to go and verify his discovery. There could be no doubt about it, and we all spent the rest of that day until dusk crawling over the gravel in the vain quest for more.'

Woodward was jubilant, for this was exactly the type of find neces-sary to destroy his rival's theories concerning the face of *Eoanthropus*; his own interpretation of the 'missing link' now being totally vindicated. In retrospect, the good fortune of the team in finding precisely what they were looking for was, of course, nothing of the sort. The discussion concerning the missing canine of *Eoanthropus dawsoni* was something that Dawson may not have envisaged when he first began the fraud in the winter of 1911/12, but, having heard and digested the expert opinion about what *needed* to be found, he set about ensuring to resolve the issue in dramatic style.

When the canine came to be scientifically examined in the early 1950s, it was found to be the that of an orang-utan which, as with the molars in the Piltdown jaw had been filed down and sandpapered smooth. Interestingly, its chocolate brown appearance was not a product of the 'ferruginous' coating that Dawson had created for the cranium, mandible, flint and faunal remains by dipping the artefacts in a solution of bi chromate, but a 'tough, flexible paint-like substance, insoluble in the common organic solvents, and with only a small ash-content'. Beneath this painted surface, the 'extreme whiteness of the dentine' demonstrated to Joseph Weiner and his team the 'essential modernity of the canine'.

Soaking in a 'solution of bi-chromate of potash' was not the only way of introducing artificial colour to specimens, which Dawson later 'confessed' to. On 23 April 1913, for example, the solicitor wrote to Woodward, casually observing that he had been experimenting with the use of colour in order to apply it to a cast of the Piltdown skull. 'I have done this in water-colour,' he noted, adding, 'which rubs off easily' perhaps to allay Woodward's concerns. 'I have got three different shades,' he went on, 'for the real, the restored, and the hypothetical. It does not look as if it had ever been restored.'

The use of paint to generate realistic shades of brown to the plaster cast of the Piltdown skull assumes a rather different light when one looks at the canine recovered by Teilhard de Chardin. Unlike the bulk of the fossil remains discovered at the site, the canine had, according to Weiner, been artificially stained with 'bitumen earth containing iron oxide, in all probability the well-known paint – Vandyke brown'. Furthermore, Weiner later noted, 'the reddish brown stain on the occlusal or chewing surface (like that on the molars) is probably also a ferruginous earth pigment applied as an oil paint (e.g. red sienna)'. The differential type of staining applied to the canine suggests a sudden change in strategy of the forger. This may be due to the fact that staining fresh teeth, of which the Piltdown canine was evidently one, was not easy. In order to get his freshly ground down tooth to appear fossilised and gravel-stained (at least to the same degree as the rest of the bone assemblage), Dawson had probably no choice other than to use paint.

It was a gamble, but the importance of the find necessitated the use of such a blatant and potentially detectable short cut.

Suspicion would inevitably have fallen upon Dawson if it were he who discovered the elusive canine of *Eoanthropus*, his eagle eyes having already spotted the faked human nasal bones in the disturbed gravel earlier in August 1912. For plausibility's sake, the special artefact needed to be found by another member of the team, preferably one who not only possessed an unimpeachable academic record, but one whom Dawson had fooled before. Venus Hargreaves, the site labourer, had not turned up for work on the Saturday of the find. Whether this was part of Dawson's plan, or whether he merely took advantage of the Hargreaves' absence, we do not know, but 30 August proved to be the optimum time for the planting of the next artefact in the hoax. When Teilhard de Chardin was moved away from the pit for a period of rest, the canine was already waiting for him on the top of a 'rain-washed' spread of excavated gravel.

Placing material in the ground for a colleague to 'discover' was potentially dangerous, not only because there was no guarantee that another member of the team would definitely find the piece (meaning that all the hard work in manufacturing it would have gone to waste), but also because every shovel-load of gravel was being carefully scrutinised and there was the very real possibility of detection. Placing the tooth onto an area of already excavated material, laid out on the ground for examination, was a masterstroke: no one would see Dawson drop the piece, whilst the sudden emergence of the canine could be easily explained as a consequence of rain washing away any covering of sediment.

No one appeared, at least in public, to be at all surprised that the new tooth *exactly* matched the canine predicted by Woodward in the original plaster cast of *Eoanthropus*; nor did anyone comment on the fortuitous way in which it made its appearance only *after* various assorted experts had decided what it ought look like. Dawson had evidently found it useful to wait until the authorities had agreed on what he and Woodward needed to find, rather than producing evidence first. The tooth was another dramatic find from the Piltdown gravels, albeit one which confirmed Woodward's hypothesis at the expense of his critics.

The canine provided an incredible finale to the 1913 season. Woodward, though ecstatic over the find, planned to hold back from a public announcement until the September meeting of the British Academy in Birmingham. Unfortunately for him, history was to repeat itself. On Tuesday

2 September the story was leaked to the press, this time to the *Daily Express*. Almost grudgingly, Woodward confirmed the next day to a reporter from the *Express* that a find of 'tremendous importance' had indeed been made, but the exact details were left vague, presumably in order to keep something back for the meeting in Birmingham. On seeing the news report, Dawson wrote immediately to Woodward noting his anger over the leak, commenting that 'the worst of it is that I have no doubt it was done by someone here who ought to have known better'. Despite his outward annoyance, however, it was more than likely Dawson himself who leaked the report, the appearance of which ensured a packed audience for the presentation.

During season three, in late June 1914, came a most sensational discovery, one that would rival anything that had been found before; something that proved *Eoanthropus dawsoni* had been a thinking, rational being. The circumstances of this new discovery were later outlined by Woodward in his book *The Earliest Englishman*: 'I was watching the workman, who was using a broad pick (or mattock) when I saw some small splinters of bone scattered by a blow. I stopped his work, and searching the spot with my hands, pulled out a heavy blade of bone of which he had damaged the end.' Washing the artefact, Woodward noted that it appeared to have been deliberately shaped into something that bizarrely looked 'rather like the end of a cricket-bat'. Dawson quickly cleared the gravel away and retrieved the rest of the bone.

More precise contextual detail concerning the artefact was supplied by Dawson and Woodward during their December 1914 presentation to the Geological Society where they noted that, when found, the bone had being lying in 'dark vegetable soil', beneath the modern hedge. The object had, therefore, not been lying in situ when 'the workman' (presumably Venus Hargreaves) had struck it with his mattock, exposing it to the air. Dawson and Woodward noted that the bone had been embedded in soil, probably redeposited during a recent period of gravel extraction. As to how it had ended up in the dark vegetable soil, Woodward hypothesised that it had been 'almost certainly thrown there by the workmen with the other useless debris when they were digging gravel from the adjacent hole'.

The overall condition of the bone implement was described as being 'much mineralised with oxide or iron, at least on the surface', the 'cut facets being slightly darker than the rest'. The artefact appeared to have been manufactured and shaped 'entirely by cutting', examination of the cut facets leaving no doubt in the mind of the team that working of the bone had been undertaken whilst still fresh. The raw material had, in all probability, derived from an elephant, smaller pieces being supplied by the striking of a thighbone on the outer edge. In turn, this suggested that *Eoanthropus* 'may, indeed, have been breaking up the bone to feast on the marrow', putting the bone fragment to one side 'because it suited a purpose which he had in mind'. Quite what this mysterious 'purpose' was, neither Dawson nor Woodward could elaborate on.

The terminal ends of the artefact provided no indicative marks of rubbing or grinding, which could provide a clue to how the piece had originally been used, only 'a slight battering' at the pointed end. 'Its shape is unique,' Dawson and Woodward were later to admit, 'and an instrument with a point would be serviceable for many purposes.' Later, when compiling data for his book *The Earliest Englishman*, Woodward seemed no closer to resolving how *Eoanthropus* may have used the bone tool other than it may have been used as a digging stick 'for grubbing up roots for food!'. Function aside, the retrieval of a tool manufactured from a large elephant bone at Piltdown had significant implications as to the perceived date of *Eoanthropus dawsoni*, the excavation team suggesting that whoever had made it was clearly a contemporary of 'an elephant bigger than the Mammoth (*Elephas primigenius*)'.

Fortunately for Dawson, neither Woodward nor any other members of the investigation team examined the elephant bone artefact closely; such detailed inspection only occurring in the early 1950s when the full extent of the Piltdown forgery came to light. Just as the Bulverhythe hammer, 'found' by Dawson on the beach between St Leonards and Bexhill, the Piltdown elephant bone had in fact been shaped with a steel chisel or knife, the sharp-edged facets of the metal blade being clearly visible along the pointed end of the artefact. This bone had not been shaped whilst relatively fresh, as Dawson and Woodward inferred in their 1914 paper to the Geological Society, but when it was already fossilised.

Serious doubts concerning the veracity of the piece had been raised before, most notability the 1914 meeting of the Geological Society in London where Dawson and Woodward first presented the artefact to the academic world. In the questions that followed the paper, W. Dale pointed out that, in his opinion, the tool marks recorded on the object appeared to have been made 'not with a flint flake, but with some stronger cutting or chopping implement', Reginald Smith noting that 'the possibility of the bone having been found and whittled in recent times must be considered'. Worse, from the point of view of the excavation team, A.S. Kennard observed that the difference between the natural surface of the bone and the cut portion implied that 'the bone was not in a fresh state when cut'. F.P. Mennell, who explained that he had experience with elephants commented that their bones tended to 'weather rapidly as soon as the flesh had decayed away'. As a consequence, whilst there was no difficulty in detaching pieces from such bones, 'they were usually so splintery and even friable, that they were unsuitable for any kind of serviceable implement'. Dawson and Woodward listened on with, we may imagine, some irritation and concern. Unfortunately, their response to this questioning is not known.

This leaves only the purpose, nature and design of the bone implement to be considered. The way in which the artefact had been shaped, with a steel-bladed tool, certainly mirrored the relatively crude design of the Bulverhythe antler hammer that Dawson had created before 1905, but that had been a relatively minor find in the solicitor's antiquarian collection. The Piltdown elephant bone was to face the attention of both the scientific and media world: surely Dawson could not expect that such a crude fraud would in any way advance his plans for *Eoanthropus*? True, the implement had been carved from the bone of an extinct form of elephant, providing a secure association for the earliest Englishman whilst also demonstrating that he was both a hunter and a tool maker, but the way in which it had been carved, the circumstances of its discovery and the overall shape of the piece (not to say all the doubts raised by members of the Geological Society) all generate a significant element of doubt as to the purpose of this particular hoax. Surely it was all a gamble too far?

It was, of course, Woodward who had first spotted the piece: splinters of bone being scattered into the air as the artefact was violently struck by the blade of Hargreaves' mattock, Dawson jumping into the pit, his fingers grubbing in the soil in order to retrieve the rest of the bone, whilst also conveniently obliterating all evidence of context and associations. Again, the object in question had not been lying in situ within the gravel, but 'about a foot below the surface, in dark vegetable soil, beneath the hedge'. This detail is critical for, although the size of the object (even when broken into two pieces) ensured that it could not easily have been transported onto site in a bag or jacket pocket without attracting the attention of another member of the team, its surroundings, the loose dark earth, meant that it could have been safely inserted into the soil without arousing suspicion that the ground had been disturbed in recent times. The bluff was, in fact, further helped by the observation that 'the soil left not the slightest stain on the specimen, which was covered with firmly-adherent pale-yellow sandy clay, closely similar to that of the flint-bearing layer at the bottom of the gravel'. It was likely, therefore, that the implement could not have been at its place of discovery for very long.

Woodward and Dawson publicly affirmed their belief that the piece had initially been disturbed by the first workmen in the gravel pit and, as an unwanted find, had been thrown out against the hedge. This was, of course, the perfect place for the artefact to be found, sealed enough below the level of hedge, but already disturbed enough so as to deflect attention away from the original context. Curiously, no one thought to ask why, if the workmen had initially been offered payment by Dawson for every artefact recovered, did they not spot an elephant femur, especially one cut into such a strange and unusual shape? Why did they toss it aside and bury it beneath a hedge? Why, also, did Dawson take so long to make anything of the artefact and even then, in the cold, hard light of day, was he so vague and reticent to offer any theories as to its possible function? Could it be that the bone tool was not of his making?

The worked elephant femur discovered close to the end of the 1914 excavation season was the last daring hoax of the Barkham Manor investigations; it was also the one to elicit most suspicion and disbelief from the members of the Geological Society to whom it was presented. This was primarily due to the fact that the object looked, uncomfortably, like a cricket bat. 'Its shape is unique,' Dawson and Woodward admitted rather lamely in their presentation, being an object 'serviceable for many purposes'. Others disagreed, Reginald Smith noting that although experimental work could conceivably demonstrate 'for what purpose it had been made', he admitted that he 'could not imagine any use for an implement that looked like part of a cricket-bat'. Having set his concerns on record, Smith ended with the curious, and perhaps rather mischievous comment: 'The discoverers were to be congratulated on providing a new and interesting problem, such as would eventually provoke an ingenious solution.'

Perhaps the reason that the elephant tool looked so much like a cricket bat was that it had not been conceived by Charles Dawson as part of the Piltdown fraud. If he had wanted to demonstrate the intelligence and skills of *Eoanthropus*, he need only point to the worked flint that had already been recovered. If he had wished to emphasise the hunting skills of the early human, he need only turn to the scattered bones of exotic animals retrieved from the pit. The elephant bone was, perhaps, a hoax too far. Could it be that someone, who knew or guessed that Dawson was in some way 'salting the mine' and was frustrated by the way in which the 'great discovery' was being uncritically accepted, decided to create something that simply could not be taken seriously? Hence the earliest Englishman was found with the earliest evidence of a national game: a prehistoric cricket bat? It is an intriguing possibility and one we shall explore in a later chapter.

The skills of the Piltdown forger have, in recent years, been elevated by some writers to near mythical status. Francis Postlethwaite, Dawson's adopted son, observed testily in a letter to the London *Times* in 1953 refuting Dawson's alleged involvement, that the solicitor could not have 'had the knowledge and the skill to break an ape's jawbone in exactly the right place, to pare the teeth to ensure a perfect fit to the upper skull and to disguise the whole in such a manner as to deceive his partner, a scientist of international repute'. The belief that the hoaxer must have been an international expert in many fields, a man combining sharply honed academic capabilities and practical skills, is, when one re-examines the evidence, fundamentally flawed.

Though the forger may appear at first glance to have possessed expert abilities in palaeontology, archaeology, geology and anthropology, this would, in Joseph Weiner's words, 'be an uncritical and exaggerated assessment of his qualities'. Yes, he required palaeontological knowledge, background and/or training, but Dawson had all this. His long experience working on the Sussex coast and Weald, sometimes in association with Samuel Beckles, combined with his voracious passion for research and reading, meant that Dawson well understood the 'significance of an apparently Lower Pleistocene or Upper Pliocene gravel deposit'. He knew what experts would expect to find in such a deposit; his own experience would tell him 'with what kind of animal fauna to stock this horizon'; he was aware of most up to date research in the field of human anatomy and evolution; and he knew what flint tools one would likely expect to be associated with such remains.

Chemical treatment and geological knowledge aside, would Dawson have been able to successfully 'plant' his fraudulent artefacts at Barkham Manor without arousing the suspicion of his colleagues and associates: did he possess the opportunity? The answer here is quite simply: 'yes'. If no one is suspecting a hoax, it is doubtful that anyone would have had the presence of mind to question the authenticity of any 'discoveries'. Given the diverse range of finds made over the three-year period between 1912 and 1915, we must plausibly assume that the hoaxer was present at all times to ensure that any given object was securely placed and reasonably recovered. As suspicion would have been aroused if only one person had discovered all of the important bone, tooth or flint artefacts at Piltdown, Dawson would need to make sure that any given piece was set down where it might reasonably be 'found' by another member of the team. If this failed, leaving the forger no option but to 'find' the artefact himself, then he would need, as in the case of the nasal bones, to ensure that sufficient independent witnesses were present in order to verify the authenticity of the find.

In all this frantic activity, Dawson's presence at the pit would never have been questioned. As discoverer and co-director of the project, as well as friends of both the landowner and tenant (not to say steward of the manor of Barkham) he could come and go as he pleased. Certainly, no unannounced appearance at the pit would occasion alarm or spark any concerns from the tenant, Robert Kenward. His office in Uckfield was only a mere hour's walk away and Dawson was often seen in the vicinity conducting both business and his antiquarian hobby.

And the motive? By 1915, *Eoanthropus dawsoni* had become a world wide celebrity. No less than three books, each one placing Piltdown Man centre stage, appeared in print: *The Antiquity of Man* by Arthur Keith, *Ancient Hunters* by William Sollas, and *Diversions of a Naturalist* by Ray Lankester. A fourth book, *The Earliest Englishman*, authored by none other than Arthur Smith Woodward, was well into the planning stage, some chapters apparently having been completed, before the project was shelved (it finally appeared in 1948, four years after Woodward's death). The year 1915 also saw an oil painting depicting the 'main protagonists' of the Piltdown discovery and subsequent debate; entitled 'A Discussion of the Piltdown Skull', it was unveiled at the Royal Academy in London.

The 'Discussion', painted by John Cooke, was an artistic interpretation of a meeting held on 11 August 1913 at the Royal College of Surgeons. To the left of centre is a reconstruction of the Piltdown skull, lying on the table with comparative human and chimpanzee remains surrounding it. Arthur Keith, conservator of the Hunterian Museum at the Royal College of Surgeons, sits impassively before the skull in a laboratory coat, callipers in hand. Behind his right shoulder stands Grafton Elliot Smith, Professor of Anatomy at Manchester University, his right arm extended, hand pointing to the cranium of *Eoanthropus* as if suggesting an amendment in anatomical detail. To Smith's right hovers Arthur Swayne Underwood (seated), professor of dental surgery at King's College, London and (standing)

Frank Orwell Barlow, technical assistant in the Geology Department of the Natural History Museum. Sat to Keith's immediate left, craning to get a slightly better view, are William Plane Pycraft, osteologist in the Department of Zoology at the Natural History Museum and Edwin Ray Lankester, formerly Director of the Natural History Museum and Keeper of Zoology. Standing behind Pycraft and Lankester, to Keith's left, are Charles Dawson and Arthur Smith Woodward, then Keeper of Geology at the Natural History Museum.

Woodward and Dawson appear united and calm, standing behind and slightly detached from the main academic debate, watching the unfolding proceedings with mild, almost paternal interest. They are the proud, elder statesmen of the Piltdown story. The positioning of both men takes further significance when one considers what lies just behind the tableaux, for there, on the back wall of the studio, hangs a framed portrait of Charles Darwin, the father of evolutionary science, facing forward towards the group. The way in which Dawson is positioned in the picture means that his features directly impinge upon the background portrait. It is almost as if, by discovering Piltdown Man, he is being posthumously acknowledged by Darwin as his ultimate heir and successor.

Although the protagonists in John Cooke's 1915 painting, 'A Discussion' are neatly divided into two opposing camps, facing each other across Keith's scientific analysis of the Piltdown skull, the overall effect is one of undeniable unity. To those who viewed the painting, it must have felt that all the scientists engaged in the debate were, with minor arguments aside, in general agreement as to the nature, interpretation and significance of *Eoanthropus dawsoni*. Unfortunately, as with most things in life, things were not that simple. In fact, 1915 was not a good year for Dawson's Man of the Dawn, for he was now to come under sustained attack from the United States of America.

The Ape Men

Late in 1915, Gerrit S. Miller Jnr, a mammalian expert at the United States National Museum of Natural History (Smithsonian Institution), published a study on 'The Jaw of Piltdown Man' in the pages of the *Smithsonian Miscellaneous Collections*. The article provided a detailed, comparative analysis of chimpanzee, gorilla and orang-utan mandibles. Miller observed that his work convinced him that, 'on the basis of the evidence furnished by the Piltdown fossils, and by the characteristics of all the men, apes, and monkeys now known, a single individual cannot be supposed to have carried this jaw and skull'. The mandible and cranium of *Eoanthropus dawsoni* were, in Miller's view, totally incompatible with one another; they came from different animals entirely.

Whilst agreeing that the deposition of such a cranium and jaw in close spatial association did raise some interesting contextual issues, Miller concluded that until further material came to light, it seemed 'proper to treat the case as a purely zoological problem by referring each set of fragments to the genus which its characteristics demand'. Miller, therefore, proposed the restriction of the name *Eoanthropus dawsoni* solely to the cranium (which itself could perhaps more plausibly be reassigned to the genus *Homo*), whilst the ape-like lower jaw suggested the presence of a 'British Pleistocene chimpanzee', which should be accorded the name *Pan vetus*. A number of prominent American and British scientists agreed with Miller's arguments, their new heresy becoming known as 'dualism' (in that they believed that *Eoanthropus* represented a composite of two very different animals). Arthur Woodward, champion of the 'monoist' view, quite naturally did not agree. In December 1915, he wrote to Charles Dawson with some feeling concerning this new perspective on his cherished Dawn Man, commenting 'I am surprised that the Smithsonian will publish such nonsense'.

In reality, Woodward and Dawson were watching the new debate over the perceived authenticity of *Eoanthropus* with some interest, for both men already possessed evidence that the association of human cranium and ape-like mandible from Piltdown was by no means a 'one off'.

Few people are today aware that the remains of Piltdown Man recovered from Barkham Manor were in fact only one of three separate finds of *Eoanthropus dawsoni* recovered in the immediate area of Piltdown prior to 1916. The two additional discoveries, characterised here as 'Barcombe Mills Man' and 'Sheffield Park Man', are usually not discussed in any great detail when the circumstances surrounding Piltdown are considered, though their relationship to the 'great hoax' are pivotal.

On 3 July 1913, Charles Dawson wrote a letter to his friend Arthur Woodward claiming to have made another spectacular discovery at Piltdown. The letter stated that Dawson had found nothing less than fragments of a second prehistoric human, but this time the remains were not from Barkham Manor, but from a site nearby. 'I have,' Dawson related to his friend, 'picked up the frontal part of a human skull this evening on a ploughed field covered with flint gravel.' Having excited Woodward with the news of the find, Dawson reverted to type, failing to provide a detailed context for the discovery, noting only that it was 'a new place, a long way from Piltdown' where 'the gravel lies 50 feet below level of Piltdown, and about 40 to 50 feet above the present river Ouse'. Dawson added, by way of an explanation for such vagary, that 'it was coming on dark and raining when I left the place but I have marked the spot'. With regard to the find itself, Dawson noted that 'it is *not* a *thick* skull but it may be a descendant of *Eoanthropus*. The brow ridge is slight at the edge, but full and prominent over the nose.'

Quite what Woodward made of this information is not known. According to Dawson's own correspondence, Woodward had a chance to inspect the skull on a visit he made to Uckfield on Friday 4 July, though Dawson pre-warned him with the words 'don't expect anything sensational'. Certainly, he and Dawson must have discussed the new find at length, and Woodward had more than enough time to inspect the general area of the alleged find during the 1913 and 1914 excavation seasons at Barkham Manor. Strange, then, that the discovery of a human cranium so close to Piltdown both geographically and morphologically, was never reported upon by either Dawson or Woodward, though both could easily have noted it at any one of their Geological Society presentations. Beyond Dawson's letter to Woodward, we do not know how he felt about the presence of human remains at this new locale, or indeed whether he ever conducted further fieldwork there.

The details concerning the provenance of the new discoveries and the information concerning their discovery are vague, even by Dawson's standards. It is, however, assumed that, though Dawson does not name the site in his letter of July 1913, the finds were probably made at or close to Barcombe Mills. The reason for believing this is that, following Dawson's death in 1916, Arthur Woodward specifically asked his widow Helene if he may examine all skeletal material in her husband's private collection prior to the public auction. On 7 January 1917, Helene wrote to Woodward: 'I have not yet come across pieces of skull answering to your description but as I am putting everything of that nature into a cupboard you will have a wide assortment from which to choose.'

There was evidently something in the collection that bothered Woodward and which he particularly needed to acquire. In late January 1917, a selection of bones derived from Dawson's collection was delivered by Frederick DuCane Goodman to Woodward's London office, apparently at the request of Helene. This selection, which was mostly cranial, comprised 'a large part of the frontal bone, a fragment of what may have been part of a right parietal, a pair of zygomatic bones which do not in any way fit the frontal, and a mandibular right second molar tooth'. Woodward, who compiled the accession details for the Natural History Museum, noted that the material had derived from the 'Pleistocene gravel in field on top of hill above Barcombe Mills railway station'. In addition to the fragments of skull was a single 'chimpanzee-like' canine, which had apparently been found in 1915. Woodward was uncertain as to the derivation of the piece, noting in the accessions register that it had probably come from the same place as the skull, though he added the proviso '(not certain)'.

Information concerning the provenance of these pieces, vague though it is, can only have been supplied by Charles Dawson, possibly as a note accompanying the artefacts. From the surviving correspondence, it would appear possible that Woodward and Dawson visited the Barcombe Mills site together as early as late April 1913. Nowhere, unfortunately, is there any specific information as to the exact place where the Barcombe Mills bone was found and under what circumstances, the 'spot' marked by Dawson never being accurately plotted.

Beyond the scanty detail in the accessions register of the Natural History Museum, Woodward never made a formal report on the Barcombe Mills assemblage. Given the apparent urgency with which he had retrieved the cranial fragments from Helene Dawson early in 1917, it does appear somewhat strange that he chose not to publicly disclose the find. Perhaps, given a preliminary analysis of the bone, he came to the decision that this was not after all part of another *Eoanthropus*. There was certainly nothing morphologically distinct about the pieces of bone (least of all their thickness), which, although they appeared to have derived from at least two individuals, appeared largely indistinguishable from that of modern humans. In fact, the only aspect of the Barcombe Mills bones which in any way resembled those from Piltdown was their chocolate brown colouration. With Woodward's enthusiasm concerning the Barcombe Mills assemblage apparently on the wane, the material was put into the stores of the Natural History Museum, where it lay forgotten.

The assemblage was finally subjected to scientific analysis in 1955 when a collective of scientists, led by Joseph Weiner, Wilfred Le Gros Clark and

Kenneth Oakley were able to confidently assert that 'all these fragments have been artificially iron-stained' whilst 'their negligibly low fluorine content and their high nitrogen content show them to be almost certainly modern'. Weiner, in his analysis of the chemical treatment of the assemblage, noted that 'one treatment involved in the staining, the bichromate, was certainly known to Dawson', the solicitor having confessed to Woodward to having experimented with a variety of staining methods such as this.

The suggestion of a second *Eoanthropus* at Barcombe Mills, some distance from Piltdown Man, would have strengthened the case for the original Barkham Manor find being genuine, something which, in 1913, was still by no means certain. Having generated the second find, it is, however, unclear why Dawson appeared to have done nothing substantive with it. That he reported the discovery to Woodward (on 3 July 1913) is clear enough, although we do not, unfortunately, know how Woodward reacted to the news. Perhaps he urged Dawson to keep the find a secret, until a better understanding of the Barcombe Mills gravels could be ascertained. Perhaps Woodward was not convinced that the Barcombe cranium was in any way similar to that of Piltdown Man and felt that its disclosure would only 'muddy the waters' of the earlier find. Perhaps he was simply not convinced about the antiquity of this new cranium.

Whatever Woodward's response, Dawson did not move any further with the Barcombe Mills discovery, and does not appear to have discussed it further either publicly or privately. Though the find was central to the whole hoax, providing the much-needed evidence for a second *Eoanthropus*, one reason for temporarily shelving it was that, by the summer of 1913, the debate surrounding Dawson's discovery had suddenly taken a different turn. The key element in the Piltdown story now seemed to surround the form taken by the missing canine that featured so prominently in Woodward's reconstruction. Dawson, therefore, changed tack, realising that the Barcombe Mills 'find' could easily be saved for later use. What mattered most at this point was the successful resolution of the 'canine debate'.

The discovery of a single 'chimpanzee-like' canine molar, possibly as late as 1915, in the same general area of the Barcombe Mills skull fragments, *may* therefore have been generated by Dawson, following the finding of the Barkham Manor canine a year before. Such a find would certainly have redrawn Woodward's attention to the alternative site at Barcombe Mills, helping to convince him of its importance. Whatever the nature of the alleged discoveries, Dawson kept them safely stored in his private collection at Castle

Lodge in Lewes. Having established *when* and roughly *where* the bone had first been found, all Dawson had to do was to wait until such time that the material was required, at which point it could be dusted off and thrust into the full glare of the academic spotlight. Dawson's death in 1916, it could be argued, abruptly terminated this particular aspect of the forgery, leaving Woodward unsure as to the true significance of the Barcombe Mills site.

The story of the 'Sheffield Park Man', or 'Piltdown II' as it came to appear in the literature, has a curiously similar echo to that of the Barcombe Mills discovery. Early in January 1915, Dawson, who had widened the area of his search for *Eoanthropus dawsoni* away from the main trenches at Barkham Manor, wrote excitedly to Woodward with news of a brand-new discovery. 'I believe we are in luck again!' he enthused, 'I have got a fragment of the left side of the frontal bone with portion of the orbit and root of nose.' Once again failing to note precisely *where* the find had derived, the solicitor said only that 'the weather has been awful'. Six months later, on 30 July 1915, Dawson wrote again to Woodward with a further tantalising reference to the new discovery. 'I have got a new molar tooth (*Eoanthropus*) with the new series,' he confided to his friend, 'but it is just the same as the others as to wear.'

Once again, Woodward's replies to Dawson's letters remain unknown, though he must have been elated. There could be no doubt about it; Dawson was referring to the new discovery of a second *Eoanthropus*, somewhere close to the original site at Barkham Manor. Whether or not Woodward recalled Dawson's earlier claims to have recovered a potential 'descendant of *Eoanthropus*' at Barcombe Mills in July 1913 is unknown, but the scientist must have felt, as with Barcombe before, to advise caution until the full nature and circumstances of the discovery could be gauged.

When Woodward was finally able to inspect and examine the new bone, at some stage after Dawson's death in 1916, all doubts concerning the genetic affiliation of the skeletal remains were dispelled. The first fragment of cranium recovered was, in his view 'exactly the same mineralised condition as the original skull of *Eoanthropus*, and deeply stained with iron-oxide'. The new piece, which appeared 'similarly thickened' provided 'a portion that was absent in the first specimen'. The smaller cranial fragment recovered by Dawson proved to be 'part of an occipital bone, which is also well fossilised, but seems to have been weathered since it was derived from the gravel'. In conclusion, Woodward felt that there was 'no reason to hesitate in referring the fragment now described to *Eoanthropus dawsoni*'.

As with Barcombe Mills before it, neither Dawson nor Woodward immediately went public with the 1915 discoveries. In retrospect, this seems strange for, unlike the assemblage retrieved from Barcombe Mills, these new pieces seemed to match the remains of *Eoanthropus* already recovered from Barkham Manor. In fact, the finding of a second 'Man of the Dawn' was precisely what both Woodward and Dawson had long been hoping for, especially as significant elements within the scientific community had recently cast doubt on the authenticity of *Eoanthropus*. Any suspicion that the human cranium and ape-like jaw from Barkham Manor had been accidentally (and unjustifiably) paired would, once and for all, be blown away by the new discovery, for the chances of such an association occurring twice were astronomical. *Eoanthropus dawsoni* was a genuine ancestor of the modern human, and the bones from Dawson's new find spot would conclusively prove it.

Woodward's most detailed description concerning both the circumstances and the location of the new site came after Dawson's death, in a formal presentation to the Geological Society on 28 February 1917 and later published in their *Quarterly Journal*. Here, Woodward noted that 'one large field, about 2 miles from the Piltdown pit, had especially attracted Mr. Dawson's attention, and he and I examined it several times without success during the spring and autumn of 1914'. Later on though, following a period of cultivation, Dawson returned to the field whereupon 'he was so fortunate as to find here two well-fossilised pieces of human skull and a molar tooth'.

It is worth pointing out that although Woodward asserts that he and Dawson examined the area of the new find 'several times', nowhere does he acknowledge that he had visited the place at, or even shortly after, Dawson's discovery of the skull and tooth. This point was later confirmed in a private letter to Ales Hrdlicka, curator of physical anthropology at the United States National Museum of Natural History in Washington, dated 26 October 1926, where Woodward, having discussed the remains of 'Piltdown II', confessed that Dawson had 'told me that he found them on the Sheffield Park Estate, but he would not tell me the exact place – I can only infer from other information that I have'.

Failure to adequately record the find spot, or communicate the provenance to Woodward, could of course be excused by the Dawson's illness, the first stages of which may have been manifest by the spring of 1915 – were it not for the fact that he had also been similarly vague at both

Barkham Manor and Barcombe Mills. As with the Barcombe Mills discovery, we cannot be sure whether the solicitor was working by himself or with others at Sheffield Park, though the implication in the letters that he sent to Woodward was that he was working alone. With Dawson as the forger working alone to create evidence for another *Eoanthropus*, the nature and details concerning the hoax become far easier to explain; there is no point looking for the location of the finds, for they were never in the ground to start with.

Doubts concerning the exact provenance of this second *Eoanthropus* meant that Woodward was, ultimately, unable to continue Dawson's examination of the alternative site in and around the Sheffield Park estate (all of his later work at Piltdown appears to have concentrated around the area of the main trenches at Barkham Manor). When Woodward finally compiled the data surrounding Piltdown Man for his book *The Earliest Englishman*, Piltdown II was confined to nothing more than a footnote, the site itself being briefly dismissed as 'a patch of gravel about two miles away from the original spot'.

In the absence of Piltdown II from the academic debate, the arguments concerning the possible dualistic nature of the human skull and ape-like jaw rumbled on through 1916 and 1917, becoming increasingly acrimonious in the process. Arthur Keith was dismissive of Miller's arguments, noting the dentition of *Eoanthropus* was 'as unlike chimpanzee teeth as teeth can be'. William Pycraft was far more damning, commenting that his conclusions were based on 'assumptions such as would never be made had he not committed the initial mistake of overlooking the fact that these remains – which, by the way, he has never seen – are of extreme antiquity, and hence are to be measured by standards of the palaeontologist rather than the anthropologist.' The failure to observe the remains from 'the right perspective' had, Pycraft observed, caused Miller to 'overlook some of the most significant features of these remains', warping his judgement in the process.

Reading Pycraft's word's, William King Gregory, vertebrate palaeontologist at the American Museum of Natural History, wrote to Miller, offering his support in defence of the disgusting 'lawyer-like, hectoring tactics' of the British osteologist. As the battle lines between dualists and monists hardened, Arthur Woodward decided it was time to bring Piltdown II, the 'Last Great' find that Charles Dawson had made at, or near, Sheffield Park, under the glare of public attention.

The presentation that Woodward made to the Geological Society on 28 February 1917 was less triumphal than it could have been, probably because of the ongoing European war, but Woodward's summing-up was conclusive. Anyone who had previously doubted that the cranium and mandible recovered from Barkham Manor were in any way connected, must, it was now abundantly clear, eat humble pie. Few speakers rose to question either Woodward or Smith following their lectures, and those that did broadly agreed with the comments made. Arthur Keith, previously an opponent of Woodward's concerning the reconstruction of *Eoanthropus dawsoni*, commented that 'these further Piltdown finds established beyond any doubt that *Eoanthropus* was a very clearly-differentiated type of being' – in his opinion a truly human type.

Having conclusively proved both the existence and authenticity of *Eoanthropus dawsoni*, effectively silencing any doubters, Sheffield Park Man (or Piltdown II) quietly sank back into obscurity. Even Woodward himself preferred to limit discussion of the second *Eoanthropus*, mentioning it only once in his book *The Earliest Englishman*, and then only in vague terms. Interestingly, it was the very 'obscurity' of such a vital and important component of the *Eoanthropus* story, that first led Joseph Weiner to investigate the whole Piltdown assemblage. Later, Weiner was to record that, at the dinner of the 1953 Congress of Palaeontologists, Kenneth Oakley of the Natural History Museum, commented that there was no record as to the exact spot that Piltdown II had been found. Weiner was amazed because the second assemblage had been crucial in convincing the academic community that the first Piltdown Man was by no means an isolated phenomenon.

Subsequent detailed analysis of the Sheffield Park remains confirmed Weiner's worst fears. First, the wear pattern on the canine molar found by Dawson in July 1915 was revealed, under a binocular microscope, to have been 'finely scratched, as though by an abrasive'. Such scratching, or grinding, across the biting surface of the tooth had clearly been applied post mortem. Detailed examination of the frontal and occipital fragments of Sheffield Park Man revealed that they, like the artificially hardened bone from Barkham Manor, contained small amounts of chromate. Far more disconcerting, however, was the observation that the frontal piece of the Piltdown II cranium could, anatomically, have formed part of the *same skull* recovered from Barkham Manor, as 'in colour and in its content of nitrogen and fluorine it resembles the first occipital … Just as the isolated molar almost certainly comes from the Piltdown mandible, it seems only too likely that this frontal fragment originally belonged to the cranium of Piltdown I.'

Such a conclusion was explosive, for the investigation team were suggesting that the right frontal piece of the second *Eoanthropus* skull had originally formed part of the first. The reason that Piltdown I and II therefore appeared, as Woodward had observed, 'similarly thickened', was because they were indeed from *the same skull*. In order to test this hypothesis, a sample of the Sheffield Park cranium was submitted for radiocarbon dating in 1987. The results, processed by the Oxford Radiocarbon Accelerator Unit, provided a date range for the Piltdown II cranium of 970 ± 140 (OxA-1394), which may be calibrated at 95.4 per cent probability as between AD 750 and 1300. The statistical differences between this determination and that provided for the Barkham Manor Piltdown Man cranium in the late 1950s (AD 1210–1480) appeared to suggest that the pieces were in fact from 'two distinct individuals'. However, given that the Oxford Radiocarbon Accelerator Unit also supplied a date for the Piltdown I mandible that was similarly at odds with the sample dated in the late 1950s, the likelihood remains that both the Piltdown I and II crania were originally from the same individual: someone who had died at some point during the early Medieval period.

Prior to deposition at Barkham Manor, the skull that formed the basis for Piltdown I must, therefore, have been deliberately broken by Dawson, selected pieces being used for the first site. A single, unused fragment of the right frontal from this broken skull could then have been put to one side for later use at Sheffield Park, where it was placed with a piece of occipital bone from a second, more 'normal' (and possibly modern) skull. The inclusion of a rhinoceros tooth in the assemblage, as with the mammalian remains recovered from Barkham Manor, hinted at a Pliocene date for this second example of *Eoanthropus dawsoni*.

Woodward had quietly held on to Dawson's Sheffield Park *Eoanthropus* since he had first been alerted to the discovery in January 1915. Part of his delay in reporting was due to the circumstances of the war then raging across the globe, the complications of his job and the deteriorating health of both Dawson and Woodward's own son Cyril. Woodward was also keen to establish the nature and circumstances of the discovery for himself, before producing an official report. To this end, Woodward and Grafton Elliot Smith had spent at least two weeks in the summer of 1916 attempting to investigate Dawson's new find, conducting in the process the last 'official' season of excavations at Barkham Manor.

The 1916 season at Barkham Manor was, in the absence of Dawson who was by then gravely ill, somewhat disappointing. The overall strategy for the renewed programme of excavation is not entirely clear, Woodward noting only that the work was conducted 'round the margin of the area previously explored'. Presumably old spoil heaps were re-examined and new ones generated. Areas close to where earlier spectacular finds, such as the elephant femur of 1914, were extended in the hope of recovering additional remains. Unfortunately, as Woodward later noted in his presentation to the Geological Society in February 1917 (which in any way was more concerned with the Sheffield Park discovery made by Dawson two years before), although much work was conducted and soil carefully examined 'neither bones nor teeth were met with'.

Despite the singular failure of the 1916 programme of investigation, and the sad death of Charles Dawson in August that year, Woodward's presentation of the Piltdown II remains from Sheffield Park generated considerable renewed enthusiasm for *Eoanthropus dawsoni*. Woodward, inspired by the hope of finding more precious evidence of the earliest Englishman, continued to excavate at Piltdown for many years, even moving to Sussex following his retirement from the Natural History Museum in 1924. Sadly, no additional remains of *Eoanthropus* were ever found. When Dawson died, Piltdown Man died with him and Woodward was, in his book *The Earliest Englishman*, able to compress the results of his final twenty-one years of fieldwork at Barkham Manor into a single paragraph.

23. A posed shot of the Piltdown excavation team of 1912, comprising Arthur Smith Woodward (extreme right), 'Chipper' the goose, Venus Hargreaves (centre), Charles Dawson (seated) and Robert Kenward Jnr.

24. The team at work in the Piltdown gravel pit. Charles Dawson (extreme left holding the sieve) and Arthur Smith Woodward examine freshly excavated soil whilst Venus Hargreaves (with pickaxe) takes a moment to rest.

25. A posed shot taken on the road from Barkham Manor (the gravel pit is to the left) in the summer of 1912. Venus Hargreaves stands at the extreme left, with Arthur Smith Woodward (and 'Chipper' the goose), Charles Dawson and Robert Kenward (with his dog).

26. The gravel pit at Barkham Manor 'in flood' during the winter of 1913.

27. A souvenir postcard, entitled 'Searching for the Piltdown Man' showing Charles Dawson (with hammer and inset left) and Arthur Smith Woodward (with sieve and inset right).

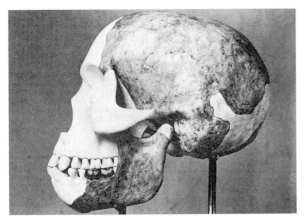

28. The plaster cast reconstruction of Piltdown Man created by Arthur Smith Woodward in 1912. The darker areas of cranium and mandible represent those pieces of skeletal remains 'found' in the first season at Barkham Manor.

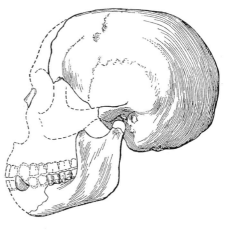

29. Arthur Smith Woodward's speculative recreation of *Eoanthropus dawsoni* (later including the nasal bones and the canine) 'found' during the 1913 season.

30. A section cut through the
gravel of Barkham Manor
in 1912 with Arthur Smith
Woodward's son Cyril standing
at the top.

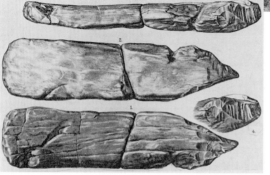

32. Four views of the Piltdown worked
bone implement.

31. Sketch compiled by Charles Dawson
to illustrate the stratigraphy of Barkham
Manor from the upper surface soil (1), the
sandy loam (2), the dark brown gravel in
which Dawson secreted the bulk of the
cultural assemblage (3), the underlying clay
and sand (4) and the undisturbed Tunbridge
Wells sand (5).

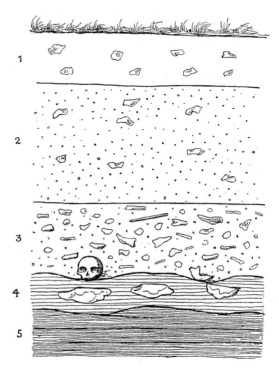

33. The 1915 painting 'A Discussion of the Piltdown Skull' by John Cooke showing, seated from left, Arthur Swayne Underwood, Arthur Keith (in lab coat), William Plane Pycraft and Edwin Ray and standing, from left, Frank Orwell Barlow, Grafton Elliot Smith, Charles Dawson and Arthur Smith Woodward. A portrait of Charles Darwin hangs on the wall in the background.

34. The artificially abraded molars of Piltdown Man from above.

35. Artificial abrasion across the surface of a molar as revealed under a scanning electron microscope.

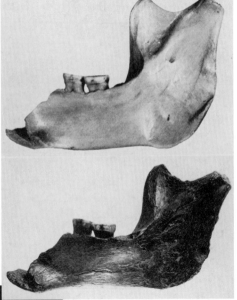

36. The mandible of *Eoanthropus dawsoni* (bottom) shown with that of a modern female orang-utan (top) broken and abraded in exactly the same way for the purposes of comparison.

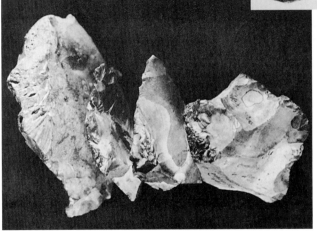

37. Four of the artificially stained 'palaeoliths' from Piltdown.

38. The worked end of the Piltdown elephant bone, showing cut marks derived from a steel knife.

39. The Morris palaeolith.

40. The Maresfield map, a clever piece of whistle-blowing drawn by John Lewis for inclusion in the *Sussex Archaeological Collections*.

41. A 'brief history of crime' as held in the archives of Hastings Museum.

42. Piltdown today.

DEALING WITH SUSPICION

Early on 10 August 1916, Charles Dawson died. He was just 52.

He died before achieving his last two goals: knighthood, an honour that others associated with the Piltdown 'discovery' and its subsequent reporting were to receive, and fellowship of the Royal Society. His sudden death also left many elements of the Piltdown forgery hanging, incomplete and unresolved. This, of course, presents anyone studying the story of the hoax with the biggest conundrum of all: had he lived, what would Charles Dawson have done next?

We know that, at the time he was taken ill, Dawson was already far advanced fabricating evidence for an additional two members of the *Eoanthropus* family, having established find spots at both Barcombe Mills and Sheffield Park. These additional frauds, designed to show that the cranium and mandible from Barkham Manor were not accidentally associated, were only in their earliest stages, both requiring more work on Dawson's behalf to establish context and associations. By the time that Dawson's fieldwork collaborator, Arthur Smith Woodward, was able to acquire the material from both these new sites, their significance was unclear, Dawson having been unable to resolve the final issues surrounding their 'discovery'. Had he lived, it is likely that Dawson would have made further finds from these areas; discoveries that proved that the Sussex Weald had been home to an entire colony of early humans.

The final years of his life, however, had seen Dawson drawing considerable attention to himself and to all of his earlier discoveries, the majority of which had been found a comparatively short distance from his home in East Sussex. Was it possible, some people were already asking, that so many incredible finds could have been made by one man in so small area of Britain? There was, by this time, considerable dislike of the solicitor in the antiquarian circles of the county, especially from within the ranks of the Sussex Archaeological Society, whom Dawson had treated with spectacular disdain in 1904, evicting them from their headquarters at Castle Lodge in Lewes. Would any more miraculous discoveries have convinced people that something was not quite right? Would a deeper investigation of Dawson's collection in Hastings Museum have demonstrated the fraudulent nature of certain pieces? Certainly, the Sussex solicitor was playing a very dangerous game: he desperately wanted to earn full recognition from the Royal Society, but any exposure of his dabblings in forgery would destroy both his reputation as an antiquarian and his career as a man of law.

Some of the writers, historians and conspiracy theorists describing the circumstances of the Piltdown hoax have suggested that Charles Dawson could not have been working on all the forgeries alone; that the skills involved in the hoaxes were way beyond those usually associated with a solicitor and amateur antiquarian. He must, some claim, have had a willing friend in the British Museum or perhaps he knew of a scientist on the take. Possibly it was an academic on the first faltering steps of their career, desperate to make a discovery that would establish their name, or perhaps it was a fellow amateur enthusiast with a grudge against the Establishment?

Dawson, however, as we have already seen, was no ordinary amateur. He was, after all, a self-professed expert in archaeology, palaeontology, photography, chemistry, biology, physics, mineralogy, anatomy, anthropology, ethnography, lithics, ceramics, metallurgy, natural history, human history, numismatics, heraldry, languages (ancient and modern), entomology and aerodynamics. He certainly possessed the skill set necessary to undertake the forgery on his own.

An accomplice is a dangerous beast. Having someone else in 'on the know' leaves the prime mover behind a hoax open for exposure or potential blackmail. True, the existence of a partner in crime would have made certain aspects of the Piltdown hoax, in particular, easier to develop, providing a distraction, allowing the other to successfully plant an artefact, or an alibi if there was the merest suspicion of foul play, but Dawson seems only to have operated alone. It was his design and his desire for glory that appears to have been the sole motive throughout nearly thirty years of fraudulent activity. He did frequently work with other amateur antiquarians, fossil hunters and scientists, it is true, but he needed to do this if the various frauds were to work. The repeated discovery of unusual or spectacular material *by himself* would have eventually aroused suspicion. Working closely with an academic dupe, allowed the solicitor to share both the glory and the blame (should anything go wrong) successfully diffusing any doubt. Having an unwitting accomplice, someone who provided an unbiased witness statement and who would later put their career and reputation on the line, either directly (in order to support Dawson) or by association (just by being there at the time the find was made), was considered important enough. Allowing them in on the joke was tantamount to academic and professional suicide.

As the Piltdown forgery began to take shape, however, significant voices of doubt were starting to make themselves heard. Had Dawson lived beyond 1916, it is possible that these voices would have started to have a major detrimental impact upon the solicitor's chances of academic advancement.

The Flint

There is a curious piece of evidence relating to the Piltdown palaeoliths, which appears to directly identify the hand of the forger. In the autumn of 1953, Joseph Weiner heard from an interviewee that a certain Harry Morris, 'a bank clerk and keen amateur archaeologist', had once believed the Barkham Manor flints to be fraudulent. Having somehow acquired one of the Piltdown flint palaeoliths, Morris had become convinced that something was wrong, voicing his concerns in a series short written statements.

Unfortunately, by the time Weiner got hold of the information, Morris was dead, although a large cabinet of flints, apparently containing the Piltdown artefact, had been bequeathed to his friend A.P. Pollard. Pollard, not being an avid collector of such things, had exchanged the cabinet with Frederick Wood for a collection of bird eggs. Wood died shortly afterwards, but, Pollard explained to Weiner, he believed that Wood's widow still lived in the village of Ditchling, to the north of Brighton. It looked to Weiner that this was going to prove to be another frustratingly intangible lead, but Mrs Wood was finally traced and, yes, the cabinet remained in her possession.

The elusive Piltdown flint that Morris had somehow obtained from Charles Dawson proved more interesting than Weiner could ever have hoped. The artefact, which was identical in lithology to those recorded by Dawson, was 'somewhat quadrangular in outline' with some evidence of battering. Directly onto the surface of the flint, Morris had written in ink 'Stained by C. Dawson with intent to defraud (all) – H.M.'. On a small note accompanying the piece Morris had added the comment: 'Stained with permanganate of potash and exchanged by D. for my most valued specimen! – H.M.' A third note, this time written on a small piece of photographic backing, continued:

> Judging from an overhead conversation there is every reason to suppose that the canine tooth found at *Pdown was imported from France!*. I challenge the S[outh] K[ensington] Museum authorities to test the implements of the same patina as this stone which the impostor Dawson says were 'excavated from the Pit!' They will be found [to] be *white* if hydrochlorate [sic] acid be applied – H.M. Truth will out.

In pencil, across the ink-written note, Morris had added 'With C. Dawson's kind regards'.

The reference to the canine tooth having been 'imported from France' appears to be a direct accusation against Teilhard de Chardin, the man who found the piece during the 1913 excavation season, though the context of Morris' 'overhead conversation' is left infuriatingly vague. A fourth and final note read: '*Dawson's Farce*: "Let not light see my black and deep desires. The eye wink at the hand; yet let that be. Which the eye fears when it ids done – to see!" Macbeth Act I, 3.'

Weiner did not at first know what to make of these extraordinary accusations. The first issue to resolve was *when* precisely they had been written. Morris had not dated any of the comments, though their tone and content suggested that they had been compiled between 1913 and 1916. Morris mentioned that he had obtained the palaeolith by exchanging it for 'my most valued specimen' (presumably another flint), which may imply a date very early on in the investigations at Barkham Manor, as, following the involvement of Arthur Smith Woodward and the Natural History Museum in the summer of 1912, all artefacts derived from the Piltdown dig went directly to South Kensington. Whatever the exact date of their writing, it was clear that the notes predated any formal accusation against either Dawson or the authenticity of the Piltdown assemblage.

Any queries concerning the authenticity of the Morris flint (as a 'genuine fake') were dispelled when Weiner applied dilute hydrochloric to the piece. This, just as Morris intimated it would, dissolved away the yellow-brown stain 'leaving a pale yellowish or greyish white surface'. Morris' challenge on the third note to the authorities of the Natural History Museum to 'test the implements of the same patina as this stone' was accepted, hydrochloric acid similarly dissolving the artificial brown stain of the three palaeoliths in their custody. Morris had been wrong in one respect, however: the staining had been caused, not by 'permanganate of potash', but by the application of bichromate solution (as per the animal bone, cranium and mandible of *Eoanthropus*).

Three questions remained unanswered: how had Morris become aware of the fraud, why had he not disseminated his evidence to a wider audience and to whom was he writing with his plea? A.P. Pollard believed that Morris had first discovered the deception because, as an expert on the area, he would have known from the geological formation of the gravel at Piltdown that it was *not* Palaeolithic. Having obtained the necessary flint from Dawson (quite how the Uckfield solicitor was persuaded to part with such a valuable item is never established – perhaps the exchange was only at first a temporary loan?), Morris was, Weiner was to record, 'utterly convinced that some deception had been carried out'. That he only seems to have conveyed his convictions to close friends perhaps belies the fact that, by 1912, Dawson was held in great esteem by archaeologists, anthropologists, geologists and palaeontologists alike. He was, furthermore, a renowned solicitor and pillar of local society (despite falling out with the Sussex Archaeological Society). For Morris to attack Dawson by announcing that the Piltdown flints were fraudulent, would be to attack all the eminent academics and scientists who were becoming increasingly involved in the *Eoanthropus* debate.

It is possible that word of Morris' accusations reached the ears of the American vertebrate palaeontologist William King Gregory, for it is he who, in 1914, had noted that unnamed individuals had suggested that Piltdown was a deliberate hoax. It is, however, important to point out here that Morris seems only to have believed the Piltdown flints to be fraudulent; we have no evidence to suggest that he similarly thought the skull and mandible of *Eoanthropus* to be. As a whistle-blower Morris was, furthermore, in something of a quandary, for he was an ardent supporter of eoliths, the crudely broken flints that some thought to be the earliest evidence of human handiwork. Dawson, as we have already seen, had publicly (and convincingly) announced his disbelief concerning authenticity of eoliths. This may well have put Morris in a dilemma. As Weiner was to speculate:

[Morris] nourished (in private) these serious accusations of fraudulent dealing, and we may be sure that he would ardently wish to see the faked palaeoliths swept away in favour of his eoliths ... but to denounce the flint [palaeolith] implements must bring the whole of the Piltdown remains into disrepute. And the continued existence of Piltdown man was vital to Morris. For no better evidence of the human workmanship and the genuine antiquity of the 'eolith' could be imagined.

Certainly, Morris clung on to his belief that the crudely broken eoliths were a genuine product of early human endeavour, so much so that, in his latter years, he appeared a 'cranky and heterodox protagonist' of a 'half forgotten and little believed theory', a point later remembered by Arthur Keith when he recalled his view that Morris had turned sour 'because of scepticism'.

Were the eoliths recovered with the remains of *Eoanthropus* in 1912 and 1913 in any case a genuine part of the forger's master plan (designed to show the range of Piltdown Man's capabilities), or were they really naturally broken flints that just happened to be located in the gravel? The eoliths did not appear to have been artificially stained and, unlike the palaeoliths, were not affected by the application of hydrochloric acid. Dawson was largely sceptical of the artefacts from the start of the investigation, something which probably indicates they were never intended to be part of his great hoax. Even if they were, it is possible that Dawson decided to remove them from the inquiry at a very early stage. This would have had a double benefit: demonstrating his objectivity (in not accepting every artefact from the gravel as genuine) and also providing him with a useful defence against Morris, should the bank clerk ever decided to go public with his story of fraudulent palaeoliths.

Dawson must also have been aware that the loss of one of his artificially stained flint artefacts from Piltdown to Morris was something that could potentially jeopardise the whole project (which again brings us to ask, why Dawson agreed to exchange it in the first place?). If, however, Morris appeared to be thinking of using the flint in evidence against Dawson, he may well have thought twice after Dawson's extravagant paper to the Royal Anthropological Institute in February 1915. To reveal Dawson's flint then would undoubtedly have reflected badly on Morris, making it look as if he were a sore loser, seeking a form of retribution over his persecutor (as many seemed to believe anyway). Ridiculing the eoliths in such an emphatic (not to say public) way would probably also have gratified Dawson immensely, certainly his letters to Woodward after the presentation seem almost proud of how he had upset the eolithophiles.

The Map

The so-called 'Maresfield map', which first appeared as an illustration in the 1912 volume of the *Sussex Archaeological Collections*, may be our only example of serious attempt at whistle-blowing in the story of Charles Dawson's 'big discovery'. The map, which has been accepted as a copy made by Charles Dawson of a genuine eighteenth-century artefact, has more recently been described as a cheap fake; something which is 'wholly fictitious' and of no historical value whatsoever. Philip Howard, writing in the *New York Times* in 1974, was more explicit; noting that the map was akin to a 'smoking gun', which proved that Dawson was an accomplished, habitual forger. The smoking gun is, however, on closer inspection far less damning than Howard enthusiastically suggests. There is no doubt concerning the map's lack of authenticity, but its connection to Charles Dawson is, as we shall see, somewhat ambiguous.

The first issue to deal with is that of legitimacy. There is no doubt that the illustration that accompanied the 1912 article 'A notice of Maresfield Forge in 1608' is a forgery, and a crude one at that. In 1974, Lieutenant Colonel P.B.S. Andrews managed to compile a detailed list of errors, mistakes and incongruities contained within the map and which counted against the genuineness of the article. For a start, there was no provenance or title given for the map and the scale employed was hugely distorted, the left side of the map being compressed to one quarter of that on the right. The scripts used were anachronistic and inconsistently applied, Andrews noting particularly the spelling 'Hondred' as a 'pseudo-archaism' for Hundred, whilst certain nineteenth-century anachronisms were set out in 'small Roman capitals of 19th–20th century type'.

A number of named places cited on the map were also hopelessly incorrect, 'Five Wents' for example, actually having been 'Six Wents' at the time the map was supposed to have been complied. The River Ouse, shown as navigable in the map, was only cleared and opened after 1790, whilst roads shown on the 1724 map actually post-dated 1830. The position of a cutting on the road to the south of Maresfield dated from the time of the 1752 turnpike. The Mill Pond, which was shown to exactly the same dimensions as on the 6in Ordnance Survey map of 1900, used early twentieth-century map conventions for a marsh. The Powder Mill as shown was not actually attested until 1852. The depiction of the stream above Maresfield forge 'explicitly denies [the] existence of [a] hammer pond', something which would make the forge itself unusable, whilst the furnace is depicted on the highest hill with no direct access to water. All these observations, Andrews concluded, served 'to demonstrate the total absurdity of the map and its uselessness as evidence for anything historical'.

To whom, then, may we credit authorship of the Maresfield map? Andrews held short of actually naming names, other than to note that the map was supposedly 'made by C. Dawson F.S.A.', as stated in the original caption of 1912. Howard, as we have already noted, was less vague, stating that 'For the first time Dawson has been caught red-handed'. Later re-examination of the map, however, shows that things are not quite that clear cut.

Although Dawson's name is appended to the figure in the final publication, there is nothing about the illustration that explicitly suggests that he had a hand in its manufacture. In fact, an analysis of the drawing's form and style strongly suggests authorship by another, in this case by sometime Dawson collaborator (at the Lavant Caves and Hastings Castle) John Lewis. There is nothing about the illustration that irrefutably demonstrates the handiwork of Charles Dawson, only that the version of it appearing in Crake's article of 1912 had been *made* by him. Some take this to imply that the map, drawn by Lewis, had merely been 'copied photographically' by Dawson in circumstances unknown, and later published by the editor of *Sussex Archaeological Collections* to illustrate an independently written article.

This observation is worth examining in greater detail for a number of reasons. First, though Dawson was indeed an accomplished photographer, there is no reason as to why, in 1912, he should be photographically copying anything for inclusion in the *Sussex Archaeological Collections*, especially as he had dramatically fallen out with the Sussex society in or around 1904 (ostensibly over the sale of Castle Lodge in Lewes). Dawson and the society certainly seem to have severed all active links by 1912, the Sussex solicitor preferring at this time to work only with the Hastings Natural History Society and Hastings Museum.

Secondly, and perhaps more tellingly, there is no reason as to why the illustration should *appear at all* with Crake's article of 1912 as there is no explicit reference to it nor any other figure in the published text, the article having been written as if it were intended to be unillustrated. Why then, if both text and illustration are not linked, the article standing well on its own without any appended figures, did anyone feel it necessary to include the map in the first place? Furthermore, why, if Dawson only *made a copy* of a drawing made by Lewis of an original map of 1724, is Lewis not credited?

A clue to all this may be found in the involvement of John Lewis himself. Lewis was a Fellow of the Society of Antiquaries and a member of the Sussex Archaeological Society from 1892 to 1907. He appears to have been a more than competent draftsman, collaborating with Dawson on a number of archaeological projects between 1893 and 1894. Relations between the two seem to have deteriorated, however, Joseph Weiner noting that in 1911 a serious quarrel had resulted in the effective termination of their friendship. The causes of the quarrel remain unknown, though John Combridge has inferred that the dispute could have supplied a motive for Lewis to perpetrate a number of hoaxes upon Dawson as a way of perhaps undermining his academic credibility.

Certainly as a forgery, the Maresfield map does not appear very good. It is lacking in technical ability and contains so many typological, cartographic and chronological errors, that it is surprising that it took until the early 1970s to be exposed. It is almost as if, by including so many obvious anachronisms and blatant mistakes, the forger actually *wanted* to be discovered. The misspelling of the word Hundred, for example, appears to have been deliberately highlighted by the positioning of an arrow-like blot, directly pointing to the mistake. All the nineteenth-century anachronisms identified by Andrews were set out in 'small Roman capitals of 19th–20th century type', something which only served to illuminate them and further emphasise their untrustworthiness. This is something which may also be inferred by the date of 1724 given to the map, for this was also the year of the original 6in:1 mile map of the county made by Budgen, something which Howard suggests may have been 'a strident warning … to recall Budgen's well-known genuine map of Sussex of that date and compare them.'

This is not, then, the work of an adept and clever hoaxer hoping to gain significant academic kudos for discovery and detailed analysis. Dawson certainly never appears to have publicly acknowledged the map or discussed it in any shape or form, in fact, if his name had not appeared so prominently within the caption appended to the map in the article, we would probably not make any link with him at all. Admittedly, there is no record of Dawson ever denying his involvement with the illustration, or that he had ever photographically copied it for inclusion within the *Sussex Archaeological Collections*, but even if he had made such a public statement, it is unlikely that the society would have published an apology or retraction given the state of affairs in the months following Castle Lodge and the later discovery of Piltdown Man. Whatever the interpretation of this blatant forgery, it is worth reiterating that there is nothing about the map that can directly associate it with the name of Charles Dawson.

It is perhaps Piltdown itself that provides the key to resolving the mystery. As both Veryan Heal and John Combridge have noted, the name 'PILT DOWN' occupies a prominent place in the drawing, a geographical impossibility made possible by the 'compression of the scale on the left of the map, to approximately one quarter of that on the right'. Piltdown, therefore, assumes an important and obvious part of the drawing, something which gains significance when one considers the timing of publication.

The 1912 edition of the *Sussex Archaeological Collections*, containing the offending map of Maresfield forge, came out in the spring of 1913, a mere three months after the Piltdown discoveries had officially been announced to the world. If, as seems likely, the map was a last-minute addition to an existing article (hence its incompatibility and lack of clear association with the published text), the prominence of the name Piltdown, combined with the sheer quantity of obvious anachronisms, is intriguing. Add to this the authorship of the map by John Lewis (who by 1911 had fallen out with the Uckfield solicitor), and the clear attempt to link the hoax with Dawson, then something rather more devious than the careless fraud may at last be glimpsed.

Here, then, it is suggested that the Maresfield map was not just any old forgery, another one designed to fool the archaeological Establishment; rather it was a clever attempt to point the finger of suspicion at both Dawson and his latest discovery at Piltdown. The Maresfield map is, therefore, a subtle attempt at whistle-blowing. Perhaps Lewis had, during his time working with Dawson, become aware that there was something not altogether above board concerning the activities of his erstwhile colleague. The discoveries made within the Lavant Caves could first have triggered his suspicion. Perhaps Dawson's interpretation of the Hastings Castle wall shadows, followed by his purchase of Castle Lodge, generated further doubt in Lewis' mind. The unearthing of apparently early human remains at Piltdown may finally have tipped the balance against the solicitor and Lewis planned to create something which was so full of errors, that it could not possibly be accepted as being genuine.

By crediting this blatant hoax to Dawson, Lewis would presumably have hoped that more people would see the amateur antiquarian for what he was. By placing the name Piltdown so centrally in the map, as well as repeatedly using the word 'forge' throughout the drawing, Lewis was perhaps further hoping to plant seeds of doubt regarding Dawson's latest discovery in the public mind. What we have, therefore, is a map of the Maresfield *forge*ry. Salzman, as editor of the *Sussex Archaeological Collections* in 1912–13, would probably have supported Lewis' actions for we have evidence, not only of his intense dislike of Dawson, but also of his satirical wit. By 1910, Salzman had little reason to like Dawson: if the whole Castle Lodge incident had not been enough, he also believed that the solicitor had been responsible for a spoiling campaign against his candidacy for election to the Society of Antiquaries, an act itself a result of Salzman's poor review of Dawson's book the *History of Hastings Castle*. That Dawson did indeed bear a grudge was something allegedly confirmed to Salzman by John Lewis himself.

Dawson must have been shaken by the publication of the Maresfield forge map. In 1913, he was still a member of the Sussex Archaeological Society, and so would have received the annual journal containing the offending drawing. It must have been abundantly apparent to him that, not only was there significant doubt concerning his latest discovery within the society, but also that his colleague John Lewis and, by implication, Louis Salzman too, were both in on his secret. Dawson never publicly denied involvement in the map, even though the caption accompanying it clearly stated that he had *made* it, but then, why should he? To deny the map would only be drawing further attention to it and to Piltdown. Even if he had made himself heard, who in the Sussex Archaeological Society would have listened? Given the situation following the eviction of the society from Castle Lodge in 1904, it is unlikely that there would ever have been either an apology or retraction within the pages of the *Collections*.

What is really curious about the whole Maresfield map incident is that, despite its obviously fraudulent origins, no one publicly questioned its authenticity. When, in 1974, Lieutenant Colonel P.B.S. Andrews commented that the map was 'wholly fictitious', Phillip Howard declared in the pages of the *New York Times* that the missing link connecting Charles Dawson to the Piltdown hoax had finally been made. The Maresfield map was the smoking gun 'positively identifying Dawson's hand in forgery'. This, of course, appears to have been Lewis' intention all along, but the exposure occurred some sixty years too late to help explode the myth of Piltdown Man at the time.

In 1915 John Lewis resigned his prestigious position as a fellow of the Society of Antiquaries. His reasons for doing so are not recorded, though it is tempting to suggest that it was in part as a protest against the society's full and total acceptance of Dawson's 'discovery'. Despite his best efforts, *Eoanthropus dawsoni* was to survive for another thirty-eight years.

The Bone

The elephant bone implement recovered by Dawson, Woodward and Hargreaves from the gravel pit of Barkham Manor in June 1914 has already been discussed, however its significance in the story of *Eoanthropus dawsoni* is still a matter for some debate.

It is true that the piece broadly fitted the profile Dawson had established for fabrication and deceit since 1895, however the precise nature of the artefact raises certain questions, not least of all surrounding its basic shape. As many observers noted at the time, the implement not only *looked* like a cricket bat, it also possessed the same profile; with a flattened face on one side, a ridge on the reverse, a bevelled 'toe' and roughly diagonal shoulders – in fact the piece is only missing a handle. Had it been made of willow, rather than fossilised bone, the identification would have been clear enough. This was indeed England's earliest cricket bat, found alongside England's earliest man.

It seems unlikely that Charles Dawson, mastermind of so many frauds in his career, would have created so obvious a forgery, especially at such a critical point of the Piltdown project. He was not known for his comedy finds, indeed forgery was a deadly serious business for him. Dawson had, with much careful planning, already established the key elements of *Eoanthropus* and had suggested, through meticulous addition of stained, worked flints, that the missing link was an intelligent hunter and manufacturer of tools. The bone implement was, strictly speaking, unnecessary; and yet there it was.

Given that by 1914 there were in Sussex a small but significant number of voices criticising Dawson and casting doubt on the reliability of his finds, it is possible that the 'discovery' of the bone artefact was intended to be something else entirely: an academic bomb designed to shake the Piltdown project to its very core. An object that was so clearly fraudulent would cast doubt upon Dawson and all of his antiquarian collection. It is possible, therefore, that the bone was just as much a surprise find to the solicitor as it was to Woodward. Could it be that the 'find' had been planted by another for the team to discover; a joke designed to show how easy it was to create fraudulent artefacts and seed the gravel of Barkham Manor?

That may explain why the piece was found embedded in the darker soils of the site, in an area that Venus Hargreaves had been asked by the directors of the project to swiftly remove, rather than the stratigraphically secure gravel (and, having created the fraud, would Dawson have really gambled on its survival by allowing Hargreaves to excavate here with a mattock?). Put in such a way, Dawson's apparent frantic scrambling about in to the soil following the discovery may have had less to do with eradicating the precise context of the bone, allowing his fraud to pass unquestioned into the finds tray, but the results of a man unnerved; someone who realised that the game could very nearly be up. He needed to swiftly retrieve all elements of this unauthorised hoax before anyone else could see and understand its extent and nature. Someone had found him out, but whom? What, ultimately, was the purpose of the new fraud?

If the implement did not form part of the Dawson master plan, then it represents the only aspect of the whole Piltdown hoax that today remains unanswered. It could plausibly represent the handiwork of a suspicious former work colleague, such as John Lewis, who had worked with Dawson before and had certainly devised one attempt to blow the whistle on his erstwhile friend with the Maresfield map. It could also have been designed (and successfully planted) through the auspices of a disgruntled member of the Sussex Archaeological Society (there were many of them), such as Louis Salzman; getting his own back for Dawson's deceit with the stamped bricks at Pevensey or for the society's eviction from Castle Lodge.

Whatever the case, and whatever the reason for its manufacture, the elephant femur became a legitimate part of the Piltdown assemblage. Not everyone accepted it, and there would be many who felt that its presence strained credulity, but there it was: a carved bone implement of 'unknown purpose'. Dawson acknowledged the artefact, perhaps feeling relatively secure in the knowledge the joke had backfired on the rival trickster, for, once the artefact had become accepted by the academic community, it could no longer effectively be used against him. If the fabrication were ever exposed, it would now more likely damage Dawson's rival more than it would the Sussex solicitor.

POSTSCRIPT

Charles Dawson died in August 1916; his wife Helene one year later. Both are buried together in the romantically overgrown cemetery of St John Sub-Castro in Lewes, East Sussex. There is no great monument to the man who, in 1912, was one of the world's most celebrated scientific investigators: a simple headstone, a short inscription and no mention of his most famous discovery, Piltdown Man.

A short distance from the cemetery, the old Dawson family home of Castle Lodge still sits in majestic grandeur at the base of Lewes Castle. The exterior has remained largely unaltered from the time that the Dawsons were in residence, but there is no blue plaque on the wall, no brass plate announcing the identity of the former owners. Across the road from the Lodge, the Sussex Archaeological Society remain in possession of Barbican House, the property they found following their eviction from Castle Lodge in 1904. Today, the sixteenth-century, timber-framed house contains a splendid museum and library, although no trace can be found within of the society's most famous member. Lewes seems to have entirely forgotten Charles Dawson FSA, FGS. He has become a non-person, almost as if he had never existed.

Further afield, there is little evidence to support the existence of either Piltdown Man or Dawson himself. The bulk of the Piltdown 'finds' are today all discretely hidden in the archives of the Natural History Museum in London. They rarely get to see the light of day. In Sussex, the firm of Uckfield solicitors that Dawson partnered with George Hart retains the name Dawson Hart, but there is no mention of their founder. To the south, in the seaside town of Hastings, there is little evidence of Dawson's extensive antiquarian collection, amassed over many decades of archaeological and anthropological investigation. Some of the artefacts recovered from his excavations at Hastings Castle, together with the extant pieces of the St Leonards-on-Sea hoard, may be glimpsed in the small, but atmospheric, Old Town Museum. But the remainder of Dawson's finds, which include the Beauport Park statuette and the Bulverhythe hammer are held unseen, deep in the bowels of Hastings Museum and Art Gallery. The fall-out from the exposure of fraud at Piltdown in the 1950s was so great that anything associated with the solicitor appears to have been swiftly and permanently spirited away. There is no history of the man amongst the great luminaries of the town; no acknowledgement of his role in the establishment of the museum.

Even in the modern village of Piltdown itself, no indication is provided to the many visitors, travellers, tourists or passers-by as to why the village name appears to be so familiar; no explanation as to why it has seeped into popular culture as a word synonymous with frauds, hoaxes and conspiracy theories. If the story of Dawson and *Eoanthropus dawsoni* had occurred in the USA, there would, at the very least, be a theme park, museum or other acknowledgement of the discovery, but in Sussex, a century on from the first announcement of Piltdown Man to the world, there is nothing. Perhaps the British are too embarrassed by the associations with fraud and forgery; perhaps they are simply too ashamed to make a fuss.

The little sandstone monolith, set up in 1938, when Piltdown Man was still considered a genuine find, to mark the location of the gravel pit and the centre of the National Park, still stands close to Barkham Manor, in what is now private land – National Park status having been quietly revoked in the late 1950s. The witness section, cut in order to create a permanently visible 'slice' through the gravel stratigraphy, still exists next to the monolith, neglected, overgrown and partially backfilled. The hedge and the road leading from Barkham Manor stand much as they did in photographs of 1912–16, but the site itself lies quiet and forgotten. There is nothing to indicate that this was, on two separate occasions during the twentieth century, the most famous place on planet earth. Even the village pub, for many years the only accessible part of the story to proudly advertise its associations with Piltdown Man has, since 2011, shamefully reverted back to its former name The Lamb.

Here the story should, by rights, end: a sad footnote to an epic tale of deceit, forgery and fraud. Dawson, however, may still have the last word.

Early in 2012, whilst visiting Anne of Cleeves' house, a splendid little museum owned by the Sussex Archaeological Society in the Lewes suburb of Southover, I was rather taken by a decorative cast iron fireback. The artefact, designed to sit at the back of a hearth, prominently displayed the rather startling image of two protestant martyrs being burnt at the stake during the sixteenth-century religious persecutions of Queen Mary. The piece had particular importance for Sussex, as the individuals depicted were two of the Lewes martyrs, executed outside of the town in 1557. The associated caption noted that the fireback, which was probably late sixteenth or early seventeenth century, was a fine example of Wealden iron and had formerly been 'in the collection of Charles Dawson'. Nothing unusual in that, as Dawson had amassed an impressive array of iron artefacts in his lifetime, which had often been used in museum displays and exhibitions, but I smiled at seeing the name, as if greeting an old friend.

Two days later, whilst reading an article on post-Medieval ironwork, I saw a picture of the fireback again, this time with the comment that the piece possessed an uncanny resemblance to a sixteenth-century woodcut depicting, not the Lewes martyrs, but of protestants dying in the flames of an East Anglian persecution. Thinking back to the previous owner of the fireback, I couldn't help but smile again. What if …

ACKNOWLEDGEMENTS

Thank you to the Sussex Archaeological Society, Hastings Museum and Art Gallery, the Booth Museum of Natural History, for permission to reproduce images taken from my 2003 book *Piltdown Man: the Secret History of Charles Dawson*. Thank you also to all at The History Press, especially Tom Vivian and Lindsey Smith, for their help, advice and unswerving belief that this final text would be delivered in 2012. Special thanks must go to all the people who commented upon my first foray into the Piltdown story in 2003 and who helpfully supplied new information, alternative pieces of evidence and different perspectives.

I owe, most of all, an immense debt of gratitude to Bronwen, Megan and Macsen who have, once again, endured the arrival of *Eoanthropus dawsoni* with smiles and calm toleration. *Eoanthropus* was sadly rather an inconsiderate houseguest who disrupted their lives, filled the dining room with paper, refused to go shopping or do the laundry and left dirty dishes all over the kitchen. It was good having him to stay once more, but hopefully now he has at last found a better place to go.

FURTHER READING

Many books (not all of them entirely objective) have been written on the subject of the Piltdown hoax, but the best starting point for further research into the topic are probably the following:

Arthur Smith Woodward's *The Earliest Englishman*, published in 1948, is a short tome but important for being an account published by one of the two directors of fieldwork at Piltdown. Woodward attempts to make sense of both the 'discovery' and of Charles Dawson's obtuse and (at times) contradictory notes, in the unshakable belief that *Eoanthropus dawsoni* was a genuine find.

The Piltdown Forgery by Joseph Weiner, chief investigator of the fraud in the 1953, was published in 1955 and reprinted for the fiftieth anniversary in 2003 with an updated introduction by Chris Stringer. The book sets out the initial work undertaken into examining the fraud and suggests possibilities for the forger's identity. Weiner notes that Dawson could not have been innocent of the hoax, but stops short of making a definitive accusation.

In 1972, Roald Millar produced a good, solid introduction to both Piltdown and the history of fossilised human remains in *The Piltdown Men*, which places blame for the fraud squarely at the feet of Elliot Grafton Smith. Millar's 1998 follow-up, *The Piltdown Mystery: the story behind the world's greatest archaeological hoax*, claims that Marie-Joesph Pierre Teilhard de Chardin was the more likely culprit for the forgery.

Charles Blinderman's 1986 book *The Piltdown Inquest* (which blames William James Lewis Abbott) and John Evangelist Walsh's *Unravelling*

Piltdown: the science fraud of the century and its solution from 1996 (which blames Dawson) both provide entertainingly detailed accounts of the fraud and of the potential perpetrators.

Frank Spencer's 1990 book *Piltdown: A Scientific Forgery* supplies an extremely detailed look at the discussions and debates that surrounded Piltdown at the time of its 'discovery', Spencer concluding that the forger was likely to have been Arthur Keith. Spencer's supplementary tome from 1990, *The Piltdown Papers: 1908–1955* outlines the surviving contemporary correspondence.

Piltdown Man: the Secret Life of Charles Dawson, published in 2003 by the current author, was the first work to fully interrogate Charles Dawson's discoveries prior to Piltdown Man. It concluded that, of all the potential suspects in the hoax, only Dawson truly possessed the means, opportunity and motive.

There are, furthermore, an excellent series of online resources relating to the Piltdown archaeological fraud, the most important of which are:

Piltdown Man, an interactive guide to the forgery by the Natural History Museum: http://www.nhm.ac.uk/nature-online/science-of-natural-history/the-scientific-process/bad-science/piltdown-man/
    The Unmasking of Piltdown Man by the BBC: http://news.bbc.co.uk/1/shared/spl/hi/sci_nat/03/piltdown_man/html/
    An annotated bibliography of the Piltdown Man forgery, 1953–2005 by T.H. Turrittin: http://www.palarch.nl/NorthWestEurope/nwe.htm.
    Piltdown Man by Richard Harter: http://home.tiac.net/~cri_a/piltdown/piltdown.html
    The Piltdown Plot by Charles Blinderman: http://www.clarku.edu/~piltdown/pp_map.html
    Piltdown Man: Case Closed by Miles Russell: http://www.bournemouth.ac.uk/caah/landscapeandtownscapearchaeology/piltdown_man_a.html

Of the many thousands of articles written on the forgery, the following represent those directly referenced within the present work:

Allcroft, H. A. 1916 Some Earthworks of West Sussex. *Sussex Archaeological Collections* 58, 65–90.
Andrews, P. 1953 Piltdown Man, *Time and Tide* 12, 1646–7.

Andrews, P. 1974 A Fictitious Purported Historical Map, *Sussex Archaeological Collections* 112, 165–7.

Ashmore, M. 1995 Fraud by Numbers: Quantitative Rhetoric in the Piltdown Forgery Discovery, *South Atlantic Quarterly* 94, 591–618.

Baines, J. 1955 *Historic Hastings*. Cinque Port Press. St Leonards-on-sea.

Blinderman, C. 1985 The Curious Case of Nebraska Man. *Science* 85, 46–9.

Boaz, N. 1981 History of American Paleoanthropological Research on Early Hominidae, 1925–1980. *American Journal of Physical Anthropology* 56, 397–405.

Boaz, N. 1987 The Piltdown Inquest, *American Journal of Physical Anthropology* 74, 545–6.

Booher, H. 1986 Science fraud at Piltdown: The amateur and the priest, *Antioch Review* 44, 389–407.

Boswell, R. 1963 Skull–diggery at Piltdown, *Baker Street Journal* 13, 150–5.

Bowden, M. 1977 *Ape-Men – Fact or Fallacy?* Sovereign Publications. Bromley.

Bowler, P. 1986 *Theories of Human Evolution: A Century of Debate, 1844–1944*. Johns Hopkins University Press. Baltimore

Broad, W. and Wade, N. 1982 *Betrayers of the Truth*. Simon and Schuster. New York.

Burkitt, M. 1955 Obituaries of the Piltdown Remains. *Nature* 175, 569.

Carr, J. 1949 *The Life of Arthur Conan Doyle*. Harper. New York.

Chamberlain, A. 1968 The Piltdown 'forgery'. *New Scientist* 40, 516.

Chippindale, C. 1990 Piltdown: Who Dunit? Who Cares? *Science* 250, 1162–3.

Clark, W. 1955a The Exposure of the Piltdown Forgery. *Proceedings of the Royal Institution of Great Britain* 36, 138–51.

Clark, W. 1955b The Exposure of the Piltdown Forgery. *Nature* 175, 973–4.

Clements, J. 1997 Piltdown Man again. *Current Archaeology* 13, 277.

Clermont, N. 1992 On the Piltdown Joker and Accomplice: A French Connection? *Current Anthropology* 33, 587.

Cohen, D. 1965 Is there a missing link? *Science Digest* 58, 96–7.

Cole, S. 1955 *Counterfeit*. John Murray. London

Combridge, J. 1977 Beeching/Ashburnham: A Georgian Dial with Edwardian Scenic Engravings. *Antiquarian Horology* 10, 428–38.

Combridge, J. 1981 Charles Dawson and John Lewis. *Antiquity* 55, 220–2.

Costello, P. 1981a Teilhard and the Piltdown hoax, *Antiquity* 55, 58–9.

Costello, P. 1981b Piltdown Puzzle. *New Scientist* 91, 823.

Costello, P. 1985 The Piltdown hoax reconsidered. *Antiquity* 59, 167–73.

Costello, P. 1986 The Piltdown hoax: beyond the Hewitt connection. *Antiquity* 60, 145–7.

Cox, D. 1983 Piltdown debate: not so elementary. *Science* 83, 18–20.

Daniel, G. 1986 Piltdown and Professor Hewitt. *Antiquity* 60, 59–60.

Dawson, C. 1894a Neolithic Flint Weapon in a Wooden Haft. *Sussex Archaeological Collections* 39, 96–8.

Dawson, C. 1894b Ancient Boat found at Bexhill. *Sussex Archaeological Collections* 39, 161–3.

Dawson, C. 1897 Discovery of a Large Supply of Natural Gas. *Nature* 57, 150–1.

Dawson, C. 1898a On the Discovery of Natural Gas in East Sussex. *Quarterly Journal of the Geological Society of London* 54, 564–74.

Dawson, C. 1898b Ancient and Modern Dene Holes. *Geological Magazine* 5, 293–302.

Dawson, C. 1903a Sussex Ironwork and Pottery. *Sussex Archaeological Collections* 46, 1–62.

Dawson, C. 1903b Sussex Pottery: a new Classification. *Antiquary* 39, 47–9.

Dawson, C. 1905 Old Sussex Glass: its Origin and Decline. *Antiquary* 41, 8–11.

Dawson, C. 1907a Note on some Inscribed Bricks from Pevensey. *Proceedings of the Society of Antiquaries.* 21, 411–13.

Dawson, C. 1907b The Bayeux Tapestry in the hands of the Restorers. *Antiquary* 43, 253–8, 288–92.

Dawson, C. 1910 *History of Hastings Castle.* Constable. London.

Dawson, C. 1911 The Red Hills of the Essex Marshes. *Antiquary* 47, 128–32.

Dawson, C. 1913 The Piltdown Skull. *Hastings and East Sussex Naturalist,* 73–82.

Dawson, C. 1915 The Piltdown Skull. *Hastings and East Sussex Naturalist,* 182–4.

Dawson, C. and Woodward, A. 1913 On the Discovery of a Palaeolithic Human Skull. *Quarterly Journal of the Geological Society of London* 69, 117–51.

Dawson, C. and Woodward, A. 1914 Supplementary note on the Discovery of a Palaeolithic Human Skull and Mandible at Piltdown (Sussex). *Quarterly Journal of the Geological Society of London* 70, 82–93.

Dawson, C. and Woodward, A. 1915 On a Bone Implement from Piltdown. *Quarterly Journal of the Geological Society of London* 71, 144–9.

de Beer, G. 1955 Proposed Rejection of the Generic and Specific Names Published for the so-called Piltdown Man, *Bulletin of Zoological Nomenclature* 11, 171–2.

Dempster, W. 1996 Something up Dawson's sleeve? *Nature* 382, 202.

de Vries, H. and Oakley, K. 1959 Radiocarbon Dating of the Piltdown Skull and Jaw. *Nature* 184, 224–6.

Dodson, E. 1981 Was Pierre Teilhard de Chardin a Co-conspirator at Piltdown? *Teilhard Review* 16, 16–21.

Downes, R. 1956 *Charles Dawson on Trial (a study in archaeology).* Unpublished manuscript. Sussex Archaeological Society, Lewes.

Durrenberger, E. 1965 More about Holmes and the Piltdown Problem. *Baker Street Journal* 15, 28–31.

Eiseley, L. 1956 The Piltdown Forgery. *American Journal of Physical Anthropology* 14, 124–6.

Elliott, D. and Pilot, R. 1996 Baker Street meets Piltdown Man. *Baker Street Journal* 46, 13–28.

Essex, R. 1977 The Piltdown Plot: A Hoax That Grew. *Kent and Sussex Journal* July–September 1955, 94–5.

Garner-Howe, V. 1997 Piltdown. *Current Archaeology* 13, 277.

Gee, H. 1996 Box of bones 'clinches' identity of Piltdown palaeontology hoaxer. *Nature* 381, 261–2.

Gould, S. 1979 Piltdown Revisited. *Natural History* 88, 86–97.

Gould, S. 1980 The Piltdown Conspiracy. *Natural History* 89, 8–28.

Gould, S. 1983 *Hen's Teeth and Horse's Toes.* Norton and Company. New York.

Grigson, C. 1990 Missing links in the Piltdown fraud. *New Scientist* 125, 55–8.

Grigson, Caroline, 1992. 'Comments', *Current Anthropology* 33.3: 265–6.

Halstead, L. 1978 New light on the Piltdown hoax? *Nature* 276, 11–3.

Hammond, M. 1979 A Framework of Plausibility for an Anthropological Forgery: The Piltdown Case. *Anthropology* 3, 47–58.

Harrison, G. 1983 J.S. Weiner and the exposure of the Piltdown forgery. *Antiquity* 57, 46–8.

Heal, V. 1980 Further light on Charles Dawson. *Antiquity* 54, 222–5.

Holden, E. 1981 The Lavant Caves. *Sussex Archaeological Society Newsletter* 34, 244.

Hoskins, C. and Fryd, C. 1955 The Determination of Fluorine in Piltdown and Related Fossils. *Journal of Applied Chemistry* 5, 85–7.

Hrdlicka, A. 1914 The Most Ancient Skeletal Remains of Man. *Smithsonian Annual Report*, 491–519.

Hrdlicka, A. 1922 The Piltdown Jaw. *American Journal of Physical Anthropology* 5, 337–47.

Hrdlicka, A. 1930 The Skeletal Remains of Early Man. *Smithsonian Miscellaneous Collections* 83, 65–90.

Jones, M. (ed.) 1990 *Fake? The Art of Deception.* British Museum Publications. London.

Keith, A. 1913 The Piltdown Skull. *Nature* 92, 197–9.

Keith, A. 1914 The Reconstruction of Fossil Human Skulls. *Journal of the Royal Anthropological Institute* 44, 12–31.

Keith, A. 1915 *The Antiquity of Man.* Norgate. London.

Keith, A. 1931 *New Discoveries Relating to the Antiquity of Man.* Norgate. London.

Keith, A. 1938 A Resurvey of the Anatomical Features of the Piltdown Skull. *Journal of Anatomy* 75, 155–85, 234–54.

Keith, A. 1950 *An Autobiography.* Watts. London.

Langham, I. 1979 The Piltdown hoax. *Nature* 277, 170.

Langham, I. 1984 Sherlock Holmes, Circumstantial Evidence and Piltdown Man. *Physical Anthropology News* 3, 1–4.

Lankester, E. 1915 *Diversions of a Naturalist.* Methuen. London.

Lankester, E. 1921 A Remarkable Flint Implement from Piltdown. *Man* 21, 59–62.

Leaky, L. 1974 *By the Evidence (Memoirs).* Harcourt. New York.

Lowenstein, J., Molleson, T. and Washburn, S. 1982 Piltdown jaw confirmed as orang. *Nature* 299, 294.

Lower, M. 1849 Iron Works of the County of Sussex. *Sussex Archaeological Collections* 2, 169–81.

Lukas, M. and Lukas, E. 1977 *Teilhard: a Biography.* Doubleday. New York.

Lukas, M. and Lukas, E. 1983 The haunting. *Antiquity* 57, 7–11.

Lyell, C. 1863 *The Antiquity of Man.* Murray. London.

Lyne, W. 1916 The Significance of the Radiographs of the Piltdown Teeth. *Proceedings of the Royal Society of Medicine* 9, 33–6.

Marston, A. 1954 Comments on 'The Solution of the Piltdown Problem'. *Proceedings of the Royal Society of Medicine* 47, 100–2.

McCann, T. 1981 Charles Dawson and the Lavant Caves. *Sussex Archaeological Society Newsletter* 33, 234.

McCurdy, G. 1913 Significance of the Piltdown Skull. *American Journal of Science* 35, 312–20.

McNabb, J. 2006 The Lying Stones of Sussex: an Investigation into the Role of the Flint Tools in the Development of the Piltdown Forgery. *Archaeological Journal* 163, 1–41.

Miles, P. 1993 The Piltdown Man and the Norman Conquest: Working Volumes and Printer's Copy for Charles Dawson's *The History of Hastings Castle. Studies in Bibliography* 46, 357–70.

Miller, G. 1915 The Jaw of Piltdown Man. *Smithsonian miscellaneous Collections* 65, 1–31.

Miller, G. 1918 The Piltdown Jaw. *American Journal of Physical Anthropology* 1, 25–52.

Miller, G. 1920 The Piltdown Problem. *American Journal of Physical Anthropology* 3, 585–6.

Moir, J. 1914 The Piltdown Skull. *Antiquary* 50, 21–3.

Montagu, M. 1954 The Piltdown Nasal Turbinate and Bone Implements: Some Questions. *Science* 119, 884–6.

Montagu, M. 1960 *An Introduction to Physical Anthropology*, (third edition). Thomas Books. Springfield.

Musty, J. 1996 Dawson Reprieved? Piltdown and XRF. *Current Archaeology* 13, 226.

Oakley, K. 1955 Analytical Methods of Dating Bones. *Advancement of Science* 12, 3–8.

Oakley, K. 1960 Artificial Thickening of Bone and the Piltdown Skull. *Nature* 187, 174.

Oakley, K. 1976 The Piltdown problem reconsidered. *Antiquity* 50, 9–13.

Oakley, K. and Weiner, J. 1953 Chemical Examination of the Piltdown Implements. *Nature* 172, 1110.

Peacock, D. 1973 Forged brick-stamps from Pevensey. *Antiquity* 47, 138–40.

Pitts, M. 2003 Piltdown. Time to Stop the Slurs. *British Archaeology* 74, 8–12.

Rosen, D. 1968 The jilting of Athene. *New Scientist* 39, 497–500.

Smoker, B. 1997 Piltdown again. *Current Archaeology* 13, 358.

Straus, W. 1954 The Great Piltdown Hoax. *Science* 119, 265–9.

Thackeray, J. 1992 On the Piltdown Joker and Accomplice: A French Connection? *Current Anthropology* 33, 587–9.

Tobias, P. 1992 Piltdown: An Appraisal of the Case against Sir Arthur Keith. *Current Anthropology* 33, 277–93.

Tobias, P. 1993 On Piltdown: The French Connection Revisited. *Current Anthropology* 34, 65–7.

Vere, F. 1955. *The Piltdown Fantasy*. Cassell. London.

Vere, Francis 1959. *Lessons of Piltdown: A Study in Scientific Enthusiasm at Piltdown, Java and Pekin.* The Evolution Protest Movement. London.

Wade, N. 1978 Voice from the Dead Names New Suspect for Piltdown Hoax. *Science* 202, 1062.

Washburn, S. 1953 The Piltdown Hoax. *American Anthropologist* 55, 759–62.

Washburn, S. 1979 The Piltdown Hoax: Piltdown 2', *Science* 203 (March 9): 955–8.

Weiner, J. 1960 The Evolutionary Taxonomy of the Hominidae in the Light of the Piltdown Investigation. IN *Men and Cultures, Selected Papers of the Fifth International Congress of Anthropological and Ethnological Sciences, Philadelphia, September 1–9, 1956.* Ed. by A. Wallace. University of Philadelphia Press, Philadelphia, 741–52.

Weiner, J. 1979 Piltdown hoax: new light. *Nature* 277, 10.

Weiner, J. and Oakley, K. 1954 The Piltdown Fraud: Available Evidence Reviewed. *American Journal of Physical Anthropology* 12, 1–7.

Weiner, J., Oakley, K. and Clark, W. 1953 The Solution of the Piltdown Problem. *Bulletin of the British Museum (Natural History) Geology* 2, 139–46.

Weiner, J., Clark, W., Oakley, K., Claringbull, G., Hey, M., Edmunds, F., Bowie, S., Davidson, C., Fryd, C., Baynes-Cope, A., Werner, A., and Plesters, R. 1955 Further Contributions to the Solution of the Piltdown Problem. *Bulletin of the British Museum (Natural History) Geology* 2, 225–87.

Zuckerman, S. 1990 A new clue to the real Piltdown forger? *New Scientist* 128, 16

INDEX